Catholicism and Science

**Recent Titles in
Greenwood Guides to Science and Religion**

Science and Religion, 1450–1900: From Copernicus to Darwin
Richard G. Olson

Science and Religion, 400 BC to AD 1550: From Aristotle to Copernicus
Edward Grant

Science and Nonbelief
Taner Edis

Judaism and Science: A Historical Introduction
Noah J. Efron

Science and Islam
Muzaffar Iqbal

Science and Asian Spiritual Traditions
Geoffrey Redmond

Liberal Protestantism and Science
Leslie A. Muray

Evangelicals and Science
Michael Roberts

Catholicism and Science

PETER M. J. HESS AND PAUL L. ALLEN

Greenwood Guides to Science and Religion
Richard Olson, Series Editor

Greenwood Press
Westport, Connecticut • London

Library of Congress Cataloging-in-Publication Data

Hess, Peter M. J.
　　Catholicism and science / Peter M. J. Hess and Paul L. Allen.
　　　　p. cm. — (Greenwood guides to science and religion)
　　Includes bibliographical references and index.
　　ISBN 978–0–313–33190–9 (alk. paper)
　　1. Religion and science—History.　2. Catholic Church—Doctrines—History.　I. Allen, Paul L.　II. Title.
　　BX1795.S35H47　2008
　　261.5′5088282—dc22　　　　2007039200

British Library Cataloguing in Publication Data is available.

Copyright © 2008 by Peter M. J. Hess and Paul L. Allen

All rights reserved. No portion of this book may be reproduced, by any process or technique, without the express written consent of the publisher.

Library of Congress Catalog Card Number: 2007039200
ISBN: 978–0–313–33190–9

First published in 2008

Greenwood Press, 88 Post Road West, Westport, CT 06881
An imprint of Greenwood Publishing Group, Inc.
www.greenwood.com

Printed in the United States of America

The paper used in this book complies with the
Permanent Paper Standard issued by the National
Information Standards Organization (Z39.48–1984).

10　9　8　7　6　5　4　3　2　1

Contents

Series Foreword	ix
Preface	xvii
Acknowledgments	xix
Chronology of Events	xxi

Chapter 1.	Introduction to Science in the Catholic Tradition	1
	Introduction: "Catholicism" and "Science"	1
	The Heritage of the Early Church	3
	Natural Knowledge in the Patristic Era	6
	Science in the Early Middle Ages: Preserving Fragments	12
	The High Middle Ages: The Rediscovery of Aristotle and Scholastic Natural Philosophy	15
	Later Scholasticism: Exploring New Avenues	21
	Conclusion: From Late Scholasticism into Early Modernity	23
Chapter 2.	From Cosmos to Unbounded Universe: Physical Sciences from Trent to Vatican I	25
	Introduction	25
	The Reformations of the Sixteenth Century	26

	The Unmaking of the Medieval Cosmos: Copernicus Revises the Heavens	30
	The Conservative Origins of a Revolution	35
	Humanism and the Foundation of the Lyncean Academy	38
	Foundations of the Modern Worldview: Galileo	41
	Catholicism and Science in the Aftermath of Galileo	49
	Catholicism and the Physical Sciences in the Seventeenth Century	52
	The Popes and Science in the Nineteenth Century	55
	Catholicism and Physical Science: Three Models of Interaction	57
	Conclusion: A Diversity of Approaches in Cosmology and Religion	59
Chapter 3.	From the Garden of Eden to an Ancient Earth: Catholicism and the Life and Earth Sciences	61
	Introduction	61
	Natural History before Evolution	63
	Enlightenment Awakenings to the History of Earth and Life	66
	Cardinal Newman: Science in Catholic Thought at Mid-Century	71
	Early Vatican Reactions to Darwinism	73
	St. George Jackson Mivart: Evolution in Theistic Perspective	76
	Evolution and Religion: Germany and France	79
	Enthusiastic Appropriation: The Case of John Augustine Zahm	80
	Studied Rejection of Darwinism: Martin S. Brennan	83
	A Progressive View from the American Hierarchy: Bishop John L. Spalding	85
	Genetics and Mendel the Monk	86
	Conclusion: From the Garden of Eden to Mendel's Garden of Peas	88

Contents vii

Chapter 4.	Catholicism and Science at Mid-Twentieth Century	91
	Introduction	91
	Catholicism and the Natural Sciences in the 1922–1965 Period	92
	Thomism before Vatican II	94
	Papacy and Doctrine	97
	The Pontifical Academy of Sciences	103
	Georges Lemaître	104
	Teilhard de Chardin	106
	Vatican II and Science	112
	John Paul II and Conclusion	114
Chapter 5.	The Legacy of Vatican II in Cosmology and Biology	117
	Introduction	117
	Hans Küng	123
	William Stoeger and George Coyne	124
	Kenneth Miller	126
	Michael Behe	130
	Jean Ladrière	132
	Ernan McMullin	134
	John Haught	137
	Bernard Lonergan	139
	Stanley Jaki	141
	Thomas Berry, Edward Oakes, Elizabeth Johnson, and Joseph Bracken	143
	Conclusion	147
Chapter 6.	Catholicism, Neuroscience, and Genetics	149
	Introduction: Science and Ethics	149
	Bioethics, Personhood, and Life	158
	Catholicism, Emergentism, and Brain Science	165
	Catholicism, Genetic Science, and Original Sin	167
	Conclusion	170

Primary Sources	173
Glossary	213
Bibliography	219
Index	235

Series Foreword

For nearly 2,500 years, some conservative members of societies have expressed concern about the activities of those who sought to find a naturalistic explanation for natural phenomena. In 429 B.C.E., for example, the comic playwright, Aristophanes, parodied Socrates as someone who studied the phenomena of the atmosphere, turning the awe-inspiring thunder which had seemed to express the wrath of Zeus into nothing but the farting of the clouds. Such actions, Aristophanes argued, were blasphemous and would undermine all tradition, law, and custom. Among early Christian spokespersons there were some, such as Tertullian, who also criticized those who sought to understand the natural world on the grounds that they "persist in applying their studies to a vain purpose, since they indulge their curiosity on natural objects, which they ought rather [direct] to their Creator and Governor."[1]

In the twentieth century, though a general distrust of science persisted among some conservative groups, the most intense opposition was reserved for the theory of evolution by natural selection. Typical of extreme antievolution comments is the following opinion offered by Judge Braswell Dean of the Georgia Court of Appeals: "This monkey mythology of Darwin is the cause of permissiveness, promiscuity, pills, prophylactics, perversions, pregnancies, abortions, pornography, pollution, poisoning, and proliferation of crimes of all types."[2]

It can hardly be surprising that those committed to the study of natural phenomena responded to their denigrators in kind, accusing them of willful ignorance and of repressive behavior. Thus, when Galileo Galilei was warned against holding and teaching the Copernican system of astronomy as true, he wielded his brilliantly ironic pen and threw down a gauntlet to religious authorities in an introductory letter "To the Discerning Reader" at the beginning of his great *Dialogue Concerning the Two Chief World Systems*:

Several years Ago there was published in Rome a salutory edict which, in order to obviate the dangerous tendencies of our age, imposed a seasonable silence upon the Pythagorean [and Copernican] opinion that the earth moves. There were those who impudently asserted that this decree had its origin, not in judicious inquiry, but in passion none too well informed. Complaints were to be heard that advisors who were totally unskilled at astronomical observations ought not to clip the wings of reflective intellects by means of rash prohibitions.

Upon hearing such carping insolence, my zeal could not be contained.[3]

No contemporary discerning reader could have missed Galileo's anger and disdain for those he considered enemies of free scientific inquiry.

Even more bitter than Galileo was Thomas Henry Huxley, often known as "Darwin's bulldog." In 1860, after a famous confrontation with the Anglican Bishop Samuel Wilberforce, Huxley bemoaned the persecution suffered by many natural philosophers, but then he reflected that the scientists were exacting their revenge:

Extinguished theologians lie about the cradle of every science as the strangled snakes beside that of Hercules; and history records that whenever science and orthodoxy have been fairly opposed, the latter has been forced to retire from the lists, bleeding and crushed, if not annihilated; scotched if not slain.[4]

The impression left, considering these colorful complaints from both sides, is that science and religion must continually be at war with one another. That view of the relation between science and religion was reinforced by Andrew Dickson White's *A History of the Warfare of Science with Theology in Christendom*, which has seldom been out of print since it was published as a two volume work in 1896. White's views have shaped the lay understanding of science and religion interactions for more than a century, but recent and more careful scholarship has shown that confrontational stances do not represent the views of the overwhelming majority of either scientific investigators or religious figures throughout history.

One response among those who have wished to deny that conflict constitutes the most frequent relationship between science and religion is to claim that they cannot be in conflict because they address completely different human needs and therefore have nothing to do with one another. This was the position of Immanuel Kant who insisted that the world of natural phenomena, with its dependence on deterministic causality, is fundamentally disjoint from the noumenal world of human choice and morality, which constitutes the domain of religion. Much more recently, it was the position taken by Stephen Jay Gould in *Rocks of Ages: Science and Religion in the Fullness of Life*:

I... do not understand why the two enterprises should experience any conflict. Science tries to document the factual character of the natural world and to develop theories that coordinate and explain these facts. Religion, on the other hand, operates in the equally important, but utterly different realm of human purposes, meanings, and values.[5]

In order to capture the disjunction between science and religion, Gould enunciates a principle of "Non-overlapping magisterial," which he identifies as "a principle of respectful noninterference."[6]

In spite of the intense desire of those who wish to isolate science and religion from one another in order to protect the autonomy of one, the other, or both, there are many reasons to believe that theirs is ultimately an impossible task. One of the central questions addressed by many religions is what is the relationship between members of the human community and the natural world. This question is a central question addressed in "Genesis," for example. Any attempt to relate human and natural existence depends heavily on the understanding of nature that exists within a culture. So where nature is studied through scientific methods, scientific knowledge is unavoidably incorporated into religious thought. The need to understand "Genesis" in terms of the dominant understandings of nature thus gave rise to a tradition of scientifically informed commentaries on the six days of creation which constituted a major genre of Christian literature from the early days of Christianity through the Renaissance.

It is also widely understood that in relatively simple cultures—even those of early urban centers—there is a low level of cultural specialization, so economic, religious, and knowledge producing specialties are highly integrated. In Bronze Age Mesopotamia, for example, agricultural activities were governed both by knowledge of the physical conditions necessary for successful farming and by religious rituals associated with plowing, planting, irrigating, and harvesting. Thus religious practices and natural knowledge interacted in establishing the character and timing of farming activities.

Even in very complex industrial societies with high levels of specialization and division of labor, the various cultural specialties are never completely isolated from one another and they share many common values and assumptions. Given the linked nature of virtually all institutions in any culture it is the case that when either religious or scientific institutions change substantially, those changes are likely to produce pressures for change in the other. It was probably true, for example, that the attempts of

Pre-Socratic investigators of nature, with their emphasis on uniformities in the natural world and apparent examples of events systematically directed toward particular ends, made it difficult to sustain beliefs in the old Pantheon of human-like and fundamentally capricious Olympian gods. But it is equally true that the attempts to understand nature promoted a new notion of the divine—a notion that was both monotheistic and transcendent, rather than polytheistic and immanent—and a notion that focused on both justice and intellect rather than on power and passion. Thus early Greek natural philosophy undoubtedly played a role not simply in challenging, but also in transforming Greek religious sensibilities.

Transforming pressures do not always run from scientific to religious domains, moreover. During the Renaissance, there was a dramatic change among Christian intellectuals from one that focused on the contemplation of God's works to one that focused on the responsibility of the Christian for caring for his fellow humans. The active life of service to humankind, rather than the contemplative life of reflection on Gods character and works, now became the Christian ideal for many. As a consequence of this new focus on the active life, Renaissance intellectuals turned away from the then dominant Aristotelian view of science, which saw the inability of theoretical sciences to change the world as a positive virtue. They replaced this understanding with a new view of natural knowledge, promoted in the writings of men such as Johann Andreae in Germany and Francis Bacon in England that viewed natural knowledge as significant only because it gave humankind the ability to manipulate the world to improve the quality of life. Natural knowledge would henceforth be prized by many because it conferred power over the natural world. Modern science thus took on a distinctly utilitarian shape at least in part in response to religious changes.

Neither the conflict model nor the claim of disjunction, then, accurately reflect the often intense and frequently supportive interactions between religious institutions, practices, ideas, and attitudes on the one hand, and scientific institutions, practices, ideas, and attitudes on the other. Without denying the existence of tensions, the primary goal of the volumes of this series is to explore the vast domain of mutually supportive and/or transformative interactions between scientific institutions, practices, and knowledge and religious institutions, practices, and beliefs. A second goal is to offer the opportunity to make comparisons across space, time, and cultural configuration. The series will cover the entire globe, most major faith traditions, hunter-gatherer societies in Africa and Oceania as well as advanced industrial societies in the West, and the span of time from classical antiquity to the present. Each volume will focus on a particular cultural tradition, a particular faith community, a particular time period, or a particular scientific domain, so that each reader can enter the fascinating story of science and religion interactions from a familiar perspective.

Furthermore, each volume will include not only a substantial narrative or interpretive core, but also a set of primary documents which will allow the reader to explore relevant evidence, an extensive bibliography to lead the curious to reliable scholarship on the topic, and a chronology of events to help the reader keep track of the sequence of events involved and to relate them to major social and political occurrences.

So far I have used the words "science" and "religion" as if everyone knows and agrees about their meaning and as if they were equally appropriately applied across place and time. Neither of these assumptions is true. Science and religion are modern terms that reflect the way that we in the industrialized West organize our conceptual lives. Even in the modern West, what we mean by science and religion is likely to depend on our political orientation, our scholarly background, and the faith community that we belong to. Thus, for example, Marxists and Socialists tend to focus on the application of natural knowledge as the key element in defining science. According to the British Marxist scholar, Benjamin Farrington, "Science is the system of behavior by which man has acquired mastery of his environment. It has its origins in techniques... in various activities by which man keeps body and soul together. Its source is experience, its aims, practical, its *only* test, that it works."[7] Many of those who study natural knowledge in preindustrial societies are also primarily interested in knowledge as it is used and are relatively open regarding the kind of entities posited by the developers of culturally specific natural knowledge systems or "local sciences." Thus, in his *Zapotec Science: Farming and Food in the Northern Sierra of Oaxaca*, Roberto González insists that

Zapotec farmers... certainly practice science, as does any society whose members engage in subsistence activities. They hypothesize, they model problems, they experiment, they measure results, and they distribute knowledge among peers and to younger generations. But they typically proceed from markedly different premises—that is, from different conceptual bases—than their counterparts in industrialized societies.[8]

Among the "different premises" is the presumption of Zapotec scientists that unobservable spirit entities play a significant role in natural phenomena.

Those more committed to liberal pluralist society and to what anthropologists like González are inclined to identify as "cosmopolitan science," tend to focus on science as a source of objective or disinterested knowledge, disconnected from its uses. Moreover they generally reject the positing of unobservable entities that they characterize as "supernatural." Thus, in an *Amicus Curiae* brief filed in connection with the 1986 supreme court case which tested Louisiana's law requiring the teaching of creation science along with evolution, for example, seventy-two Nobel Laureates, seventeen state academies of science and seven other scientific organizations argued that

> Science is devoted to formulating and testing naturalistic explanations for natural phenomena. It is a process for systematically collecting and recording data about the physical world, then categorizing and studying the collected data in an effort to infer the principles of nature that best explain the observed phenomena. Science is not equipped to evaluate supernatural explanations for our observations; without passing judgement on the truth or falsity of supernatural explanations, science leaves their consideration to the domain of religious faith.[9]

No reference whatsoever to uses appears in this definition. And its specific unwillingness to admit speculation regarding supernatural entities into science reflects a society in which cultural specialization has proceeded much farther than in the village farming communities of southern Mexico.

In a similar way, secular anthropologists and sociologists are inclined to define the key features of religion in a very different way than members of modern Christian faith communities. Anthropologists and sociologists focus on communal rituals and practices which accompany major collective and individual events—plowing, planting, harvesting, threshing, hunting, preparation for war (or peace), birth, the achievement of manhood or womanhood, marriage (in many cultures), childbirth, and death. Moreover, they tend to see the major consequence of religious practices as the intensification of social cohesion. Many Christians, on the other hand, view the primary goal of their religion as personal salvation, viewing society as at best a supportive structure and at worst, a distraction from their own private spiritual quest.

Thus, science and religion are far from uniformly understood. Moreover, they are modern Western constructs or categories whose applicability to the temporal and spatial "other" must always be justified and must always be understood as the imposition of modern ways of structuring institutions, behaviors, and beliefs on a context in which they could not have been categories understood by the actors involved. Nonetheless it does seem to us not simply permissible, but probably necessary to use these categories at the start of any attempt to understand how actors from other times and places interacted with the natural world and with their fellow humans. It may ultimately be possible for historians and anthropologists to understand the practices of persons distant in time and/or space in terms that those persons might use. But that process must begin by likening the actions of others to those that we understand from our own experience, even if the likenesses are inexact and in need of qualification.

The editors of this series have not imposed any particular definition of science or of religion on the authors, expecting that each author will develop either explicit or implicit definitions that are appropriate to their own scholarly approaches and to the topics that they have been assigned to cover.

Richard Olson

NOTES

1. Tertullian, 1896–1903. "Ad nationes," in Peter Holmes, trans., *The Anti-Nicene Fathers*, ed. Alexander Roberts and James Donaldson, vol. 3 (New York: Charles Scribner's Sons), p. 133.

2. Christopher Toumey, *God's Own Scientists: Creationists in a Secular World* (New Brunswick, N.J.: Rutgers University Press, 1994), p. 94.

3. Galileo Galilei, *Dialogue Concerning the Two Chief World Systems: Ptolemaic and Copernican* (Berkeley and Los Angeles: University of California Press, 1953), p. 5.

4. James R. Moore, *The Post-Darwinian Controversies: A Study of the Protestant Struggle to Come to Terms with Darwin in Great Britain and America, 1870–1900* (Cambridge: Cambridge University Press, 1979), p. 60.

5. Stephen Jay Gould, *Rocks of Ages: Science and Religion in the Fullness of Life* (New York: The Ballantine Publishing Group, 1999), p. 4.

6. Ibid., p. 5.

7. Benjamin Farrington, *Greek Science* (Baltimore: Penguin Books, 1953).

8. Roberto González, *Zapotec Science: Farming and Food in the Northern Sierra of Oaxaca* (Austin: University of Texas Press, 2001), p. 3.

9. *72 Nobel Laureates, 17 State Academies of Science and Seven Other Scientific Organizations. Amicus Curiae* Brief in support of Appelles Don Aguilard et al. v. Edwin Edwards in his official capacity as Governor of Louisiana et al. (1986), p. 24.

Preface

This book addresses the relationship between the Catholic tradition and "natural philosophy"—what we now call "science"—across two millennia, from the early church to the present day. The volume is organized both chronologically and topically, discussing the historical relationship between Catholic theology and the contemporary scientific understanding of the different cultures and periods in which it has been embedded. The authors have made a judicious selection of those figures and episodes most illustrative of this long and colorful relationship.

Chapter one sketches the foundations of the Catholic tradition and its relationship to natural philosophy up to the later Middle Ages. In chapter two the authors trace—from Copernicus to the beginning of the twentieth century—the revolutions in physics and cosmology that radically and permanently altered the worldview of the Christian West. The same time period is treated in chapter three with respect to the gradual development of a sense of geological history and of the fundamental relatedness of all life, culminating in Darwin's theory of evolution. The most important figures, doctrinal developments, and controversies are surveyed from the perspective of understanding the changing relationship between Catholicism and the natural sciences. The last three chapters survey the abundant reflections, responses, and reactions within the Catholic Church to the modern sciences over the course of the twentieth century and into the twenty-first century. These chapters include a discussion of perspectives ranging from those of the leadership of the Church to the diverse views of individual Catholic scientists, philosophers, and theologians. The concluding chapter reviews current theological challenges to Catholicism from contemporary scientific discoveries. The book includes a bibliography that directs the reader to more detailed discussions of particular episodes in the history of the Catholic Church's dealings with science, and a glossary and chronology of key events.

Anticipating an audience of college or university students, we have not presupposed a great deal of historical background, nor of scientific or theological knowledge. Much of the material that is covered here can be found in lengthier specialized monographs focussing on specific periods. What those texts do not do is to survey the entire span of the relationship between Catholicism and the natural sciences. Don O'Leary's *Roman Catholicism and Modern Science* (Continuum, 2006) offers a good treatment of the Catholicism-science relationship since Galileo, but O'Leary also proposes a particular interpretation of many of these issues, which is what this book seeks to avoid. Instead, we retain a descriptive approach in which we endeavour to remain attentive to the theological dimensions of various questions and historical episodes. In addition to its use in the classroom, we anticipate that the book will be used for parish discussion groups and personal enrichment.

Acknowledgments

Paul would like to thank Monica, Jeremy and Sarah for their curiosity, love and patience as several chapters for this book came together. Many thanks also go to my colleagues in the Department of Theology (Concordia University, Montreal) as well as to the library staff who retrieved what I had thought was impossible to retrieve.

Peter would like to thank Viviane, Michael and Robert Hess for their love, understanding, and patient support in both times of contentment and hours of adversity during the long gestation of this book. He is grateful to Hamilton Hess for critiquing chapters 1–3, and to Genie Scott and her colleagues at the National Center for Science Education for their encouragement and assistance. The enthusiasm of graduate students at Saint Mary's College has reminded him that scholarship thrives best in a spirited community of discourse. He dedicates his portion of this book to his parents, Margaret E. Barnwell Hess and Hamilton A. B. Hess, D.Phil. Oxon., who opened his mind to the beauty of the world and his heart to the love of God.

Chronology of Events

ca. 399 BCE — Death of Socrates.

ca. 347 BCE — Death of Plato, student of Socrates and teacher of Aristotle.

335 BCE — Aristotle founds his Academy.

ca. 33 CE — Jesus executed.

33–100 CE — Apostolic Age: the period in early church history during which some of Jesus' original twelve apostles were still alive. The gospels and Pauline letters were written, church doctrine and structure began to take shape. This period ended at about the close of the first century AD, perhaps with the death of John the Apostle.

79 CE — Death of Pliny the Elder, Rome's great natural historian, while examining the eruption of Mt. Vesuvius.

2nd century — Claudius Ptolemy of Egypt formulates cosmological model that would endure until Copernicus. Galen of Pergamum (now in Turkey) wrote medical treatises that remained authoritative until the sixteenth century.

204–270 — Plotinus, founder of Neo-Platonism.

303–305 — Diocletian Persecution: last and most severe Roman persecution of Christian sect, killing between 3,000 and 5,000 Christians.

313	Edict of Milan: letter of Emperor Constantine declaring that the Roman Empire would be neutral with respect to worship.
325	Council of Nicea: rejects the Arian heresy and proclaims that the Son of God is of one substance with the Father.
354–430	Augustine of Hippo.
400–550	Barbarian invasions.
410	Alaric's Visigoths sack Rome.
450	Huns invade what is now France and Italy.
451	Council of Chalcedon defines the two natures of Christ.
ca. 480–547	Benedict of Nursia; founds Monastery of Monte Cassino ca. 529.
500–1000	Early Middle Ages.
570–622	Mohammed and the foundation of Islam.
1095	Pope Urban II preaches the first crusade against the Muslims holding Jerusalem.
12th century	Rediscovery of the works of Aristotle that had been preserved by Islam and Byzantium.
1215	Fourth Lateran Council—Eucharistic change defined as "transubstantiation"; council declares creation *ex nihilo* (out of nothing).
13th century	Medieval universities founded at Paris, Oxford, Bologna, Cambridge; two dozen more by the end of the thirteenth century.
1255	Aristotle's natural philosophy books added to the curriculum at Paris.
1265–1321	Dante Alighieri; *Divine Comedy* embodies the medieval unity of knowledge.
1277	Bishop Stephen Tempier issues condemnation of 219 propositions drawn mostly from the works of Aristotle. Encouraged natural philosophical speculation along non-Aristotelian lines.
1348–1350	Black Death: epidemics of Bubonic plague swept through Europe, killing up to one third of its inhabitants.

Chronology of Events

1323–1382	Nicole Oresme speculates in non-Aristotelian directions.
1337–1453	Hundred Years War: conflict fought sporadically over 116 years, driven by English claims to the French throne.
ca. 1453	Johann Gutenberg invents printing by movable type, making possible an explosion of book-based learning across Europe. Printing facilitates the Reformations and the Scientific Revolution.
1453	Ottoman Turks capture Constantinople (now Istanbul) and end the Byzantine Empire.
1469–1536	Erasmus of Rotterdam, humanist, publisher of the Greek New Testament, critic of the Church who would not join the Reformation.
1492	Columbus lands on the island of Hispaniola; "New World" enters the European consciousness.
	Ferdinand and Isabella capture Granada, concluding the reconquista, the recapture of the Iberian peninsula from the Moors.
1517	Luther's Ninety-Five Theses on Indulgences sparks the Reformation.
1530s	Reformation established; Schmalkaldic League.
1534	English Reformation begins with Henry VIII's Act of Supremacy, which severs ties with the Roman Catholic Church.
1537	Cosimo de Medici comes to power in Florence.
1543	Copernicus publishes *De revolutionibus orbium coelestium*.
1545–1563	Council of Trent.
1561–1626	Francis Bacon, philosopher, statesman, champion of the empirical scientific method, proponent of collaboration by sharing the results of research.
1562–1598	French Wars of Religion: civil wars between Catholics and Huguenots.
1564–1642	Galileo's life.
1572	Comet proves there are no crystalline spheres.
1577	Stella nova or new star: the fact that it had no parallax proves that it is a star, and that the heavens are not immutable, as Aristotle had taught.

Chronology of Events

1598	Edict of Nantes: proclaimed by King Henry IV of France; it ended the French Wars of Religion by defining the rights of Huguenots. Revocation of the edict by Louis XIV in 1685 cost France the talent of many Protestants who fled to Holland.
1603	*Accademia de Lincei* established in Rome.
1608	Galileo examines the heavens with a telescope.
1610	*Starry Messenger*: moons of Jupiter, mountains on the moon, phases of Venus.
1613	*Letter on sunspots.*
1616	First trial.
1627–1691	Robert Boyle: natural philosopher, chemist, inventor; best known for formulating Boyle's law about behavior of gases.
1632	Second trial; abjuration.
1643–1727	Isaac Newton; publishes his laws of motion in *Principia mathematica philosophia naturalis* (1687).
1660	Royal Society founded "for the improvement of natural knowledge."
1687	Newton's *Principia mathematica* (Mathematical Principles of Natural Philosophy) with his three laws of motion.
18th century	Age of Enlightenment.
1731–1802	Erasmus Darwin.
1734–1796	Development of the nebular hypothesis by Swedenborg, Kant, and Laplace.
1744–1829	Jean-Baptiste de Lamarck.
1750–1830s	Gradual discovery of the deep history of time.
1769–1832	Georges Cuvier.
1791	Immanuel Kant's *Critique of Pure Reason* (Germany) initiates his critical philosophy.
1798	Thomas Malthus publishes *An Essay on the Principle of Population.*

Chronology of Events

1809	Jean Baptiste Lamarck proposes progressive evolutionary theory in *Philosophie zoologique.*
1809–1882	Charles Darwin. *On the Origin of Species* published in 1859.
1830–1833	Charles Lyell publishes Principles of Geology, supporting Hutton's idea of an ancient earth, and establishing uniformitarianism.
1847	Lincei becomes "Academy of Sciences of the Pontifical States."
1851	Auguste Comte publishes *The System of Positive Polity*, proposing a "religion of humanity."
1851–1921	Holy Cross Priest John Augustine Zahm publishes *Evolution and Dogma.*
1858	John Henry Cardinal Newman Publishes *The Idea of a University.*
1864	*Syllabus of Errors:* controversial document of Pius IX condemning such modern themes as freedom of religion and separation of church and state.
1869–1870	Vatican Council I.
1881–1955	Pierre Teilhard de Chardin.
1893	Pope Leo XIII's encyclical Providentissimus Deus, rejecting much of modern biblical criticism.
1893	Encyclical *Providentissimus Deus*, censuring higher criticism of the bible.
1907	Encyclical *Pascendi Dominici Gregis* urges that sacred studies take priority over study of the social and natural sciences in seminaries.
1936	Reestablishment of Pontifical Academy of Sciences.
1943	Encyclical *Divino Afflante Spiritu.*
1950	Encyclical *Humani Generis* of Pius XII (some acceptance of evolution).
1951	Pius XII endorses big bang cosmology as divine creation.
1962	June 30—Vatican Holy office (now Congregation for the Doctrine of the Faith) issues monitum regarding Teilhard's writings.

1962–1965	Vatican Council II.
1968	Encyclical *Humanae Vitae* ("On Human Life") of Paul VI.
1992	Report on the Galileo crisis by the Pontifical Academy of Sciences; John Paul II establishes Pontifical Academy of Life.
1996	John Paul II Message to Pontifical Academy of Sciences on Evolution.
1998	John Paul II encyclical *Fides et Ratio*.

Chapter 1

Introduction to Science in the Catholic Tradition

INTRODUCTION: "CATHOLICISM" AND "SCIENCE"

The relationship between Catholicism and science in the modern world springs from a complex intellectual ecology, with multiple roots penetrating deep into the varied soils of Western tradition. There is no simple way of describing how the Christian religion—during different historical periods, in diverse geographical settings, and through changing political, economic, and social circumstances—came into contact with the study of the natural world. Because language is of great significance in our exploration of this history, we propose to begin with the terms themselves. How is our understanding of "Catholic" and "science" different from how these terms were understood in earlier eras of Christian history?

The Catholic tradition dates back to the Roman Empire. Not long after the apostolic age, the term "catholic church" was used for the first time to refer to the universal Christian community by Ignatius of Antioch (ca. 35–107) in his Letter to the Smyrneans:

See that you all follow the *bishop*, even as Jesus Christ does the *Father*, and the *presbytery* as you would the *apostles*; and reverence the *deacons*, as being the institution of *God* ... Wherever the *bishop* shall appear, there let the multitude [of the people] also be; even as, wherever Jesus Christ is, there is the *Catholic Church*. (Ignatius, 1979:90)

"Catholic" (like "orthodox") was used in both the Greek-speaking (eastern) and Latin-speaking (western) churches, and signified "whole" or "universal." In the western empire in the fourth century, however, the term began to take on a distinct theological meaning, following the

Diocletian Persecution (303–305), when the church faced the challenge of the Donatist schism. Bishop Donatus in the Roman province of North Africa excluded from communion those Christians who had renounced their faith under persecution, refusing to welcome them back into the fold (Tilley, 1997:162–171). His ecclesiology was rejected by the mainstream Western Church after Constantine legalized Christianity with the Edict of Milan in 313. "Catholic" came to signify the universal church—in conscious distinction from Donatism—just as "orthodox" came to stand for the mainstream church of the eastern half of the empire in distinction from the monophysite and Nestorian heresies.

By the end of the third century, the Roman Empire was both enormous and culturally diffuse, and its inherent centrifugal forces made continued cohesion difficult. Emperor Diocletian divided this cumbersome political creation into an eastern and a western empire along linguistic, geographical, and historical lines. The Christian church, while not theologically divided, began a long drift into two separate bodies, relationally separated by mutual excommunications. (Both communions would continue to regard themselves as the legitimate historical continuation of the primitive church through apostolic succession.) The Eastern Church was defined by the Greek language and temperament and modes of theological expression, and would in time become the Orthodox Church, the ecclesiastical face of the Byzantine Empire. The western or Catholic Church was defined by the Latin language and character and theological style, and by a more juridical approach to church organization and theology. With the tragic schism in 1054 between the eastern and western halves of the church (further deepened in 1453), "Catholic" took on a more particularly geographic connotation, referring to the church in lands that had formerly been part of the western Roman Empire, just as "Orthodox" came to refer to churches in Greece, the Levant, the Balkans, and Russia.

Finally, with the Protestant and Anglican Reformations in the sixteenth century, "Catholic" came to designate those Christian communities under the western umbrella of faithfulness to Rome, rather than those embracing the doctrines of Luther, Calvin, Cranmer, or the radical brethren. In this book—while acknowledging the historical particularities that have contributed to different senses of the term—the authors use the words "catholic" and "Catholicism" to denote the community of Christians distinct from both Protestantism and Eastern Orthodoxy.

A similarly complex history applies in the case of the term "science." The Latin word "scientia" originally meant knowledge or practical understanding, and was contrasted with "sapientia" or wisdom ("sophia" in Greek). In the middle Ages the subject matter of science would have been anything that can be organized and studied systematically, including theology, the "queen of the sciences." The umbrella of "natural history" included the study of animals and plants, rocks, minerals, and other curiosities; that of "natural philosophy" included such subjects as astronomy, motion, cosmology, and meteorology.

In the later Middle Ages the pace of philosophy quickened, and Copernicus' resuscitation of the heliocentric hypothesis stimulated the exploration of diverse ideas in natural philosophy, both old and new, such as the atomic theory of matter, the existence of a vacuum, and the circulation of the blood. In the seventeenth century, the "New Philosophy"—in contrast to the received tradition of Aristotelianism—came to designate the approaches and disciplines arising during the "scientific revolution." Isaac Newton published his comprehensive new theory of astronomical motion as *Philosophiae Naturalis Principia Mathematica* (*Mathematical Principles of Natural Philosophy*). More broadly, his subject was, of course, what today we would call physics. Likewise, alchemy would develop into chemistry proper in the next century. Geology would gain professional respectability, and biology, paleontology and allied sciences would branch from natural history. The term "scientist" was first employed in 1834 by William Whewell to designate a practitioner of one of these new divisions of knowledge.

THE HERITAGE OF THE EARLY CHURCH

In order to understand and appreciate the relationship between Catholicism and the sciences in the modern world, we must first review the various traditions that lie at the root of Christendom. It was as faithful Catholics that Copernicus and Galileo launched their critiques of the received Aristotelian-Ptolemaic astronomy. The Christian scholastic culture of the Middle Ages itself enjoyed a deep philosophical and religious heritage. This included the elements of Hebrew religious thought, Greek rationalism, Roman technology, and the Arabic transmission of and elaboration upon Aristotelian philosophy.

The first element to consider is the Hebrew religious tradition, out of which grew Christianity. Although the gospels of Luke and John portray Jesus theologically as universal savior and Word of God incarnate, historically we know that Jesus was a Jewish inhabitant of first-century Palestine (McClymond, 2001:356–375; Meier, 1991:205–371). The earliest Christian communities—initially Jewish but with a growing number of gentile converts after AD 50—were deeply rooted in the Hebrew scriptures, to which were gradually added the gospels and apostolic letters that form the New Testament canon. What Christians came to know as the Old Testament did not in itself offer a serviceable scientific picture of the world, and it certainly included no rational reflection upon the cosmos. Rather, the Hebrew scriptures record the developing moral relationship with God as experienced by a small, initially pastoral people in the Ancient Near East. However, it bequeathed two elements of great importance for the future of natural philosophy: first, the affirmation of creation as fundamentally good and as a time-related event (Genesis 1), and, second, the replacement of archaic patterns of circular or undirected time with a more linear sense of time unfolding according to a divine plan, or a "history of salvation."

The Genesis creation story expresses a clear relationship between God and the world. The sixfold divine affirmation of creation as "good" in Genesis 1 suggests that in Hebrew thought "revelation" did not merely designate the communication of divine truth to humankind, but in a real sense involves the gift of the natural world as a whole. The sharp distinction between God and creation that is characteristic of the monotheistic traditions ensured that creator and creation could not be conflated, and thus it opened the door to scientific investigation of the world. Of course, Hebrew authors were in no sense "scientists" offering a systematic account of the organic world around them—the mythic structure of the first eleven chapters of Genesis clearly shows that. However, what would hold great significance for the future of natural history in the West was the positive appraisal of nature found throughout Hebrew literature, exemplified strikingly by the psalms: "The heavens declare the glory of God, and the firmament shows forth his handiwork" (Psalm 19:1). If all creatures—including humans—praise God (Psalm 148) then surely the investigation of natural philosophy through the creatures of God is for us a legitimate enterprise.

If Hebrew religion supplied the moral foundation for Christianity, it was the Hellenistic Greek philosophy permeating the later Roman Empire that supplied its intellectual foundation (Lindberg, 1992:21–45). In the sixth and fifth centuries BCE, in the thought of some pre-Socratic philosophers, Greek mythology was eclipsed by rational confidence in an intelligible cosmos. This manifested itself in the schools of Pythagoras (ca. 580–500 BCE) and Plato (427–347 BCE) in a fascination with number and form. The idealization of numbers in the Platonic strains of Renaissance philosophy would play a crucial role in the mathematization of astronomy. Among Plato's major contributions to natural history was his epistemological "theory of forms," which served to undergird the concept of "species" as an entity truly existing within the divine mind, a theory that was not seriously challenged until the eighteenth century.

Aristotle's philosophy was equally foundational for the Catholic tradition. Although not a scientist in the modern sense of the word, his practical inclinations led him to investigate many aspects of the natural world and to compose books on a wide range of topics related to it (Lindberg, 1992: 47–68). Aristotle (384–322 BC) made seminal contributions in more areas than can here be enumerated, so we note only three. First, his doctrine of the soul as assuming vegetative, animate, and rational forms established a crucial principle of continuity within the biological realm, even if his assertion that the source of the rational soul comes from outside the physical process of reproductive transmission would set up later tensions by encouraging Christian theologians to assert that the soul was separately created by God. Second, his careful attention to firsthand observation—whether in his own work or in the reports of others—was conducive to what would become a tradition of empirical research resuscitated in the twelfth century and increasingly pursued until its firm establishment in the seventeenth century. Third, Aristotle's pervasive teleology ensured that final causes (or purposes in nature) would serve at least to 1700 as a fundamental organizing principle. In natural history teleology governed the understanding of how bodies are organized, of how individual organs are designed for specific purposes, and of how all living beings are connected through the medieval "great chain of being." In Aristotle's thought, religion and science were at best only tangentially (and perhaps negatively) related: the remote "Prime Mover" was apparently unconcerned with individual plants or animals or human beings. Aristotle's geocentric conception of the physical universe, updated by Ptolemy in the second century AD, would serve as the foundation for the vision of an integrated cosmos that endured through the Christian Middle Ages until it began to unravel in the sixteenth century.

A third element of the Christian heritage was the contributions from Roman culture, including pragmatism, technological precision, and a sophisticated appreciation of law and politics. Most significant among Roman sources for Christian writers on scientific topics was the work of Pliny the Elder (d. AD 79 in the eruption of Mt. Vesuvius), whose voluminous *Historia naturalis* exercised a profound influence on the Christian West well into the Renaissance. The encounter of natural philosophy and religion in Pliny's thought took place against the background of his central concern to catalogue and exhibit nature in all its strange and wonderful variety. Himself a skeptic about the existence of the gods, Pliny proceeded from the assumption that "the world is the work of nature and the embodiment of nature itself" (*Historia naturalis*, II.1). Thus, his collection of facts about everything imaginable can almost literally be said to articulate his "theology." Although eclectic and uncritical in nature, Pliny's *Natural History* would exert a far-reaching impact on the future of the science, for it was one of the most influential works to survive the collapse of Roman civilization, and early medieval encyclopedists overwhelmingly drew from it or even patterned their own works upon it.

In brief, the first Christians enjoyed a rich and multifaceted legacy of natural philosophy, which over the course of the next millennium would inform their view of the world, and in turn, their theology. We must recognize it both for what it was, and for what it was not, as noted by David Lindberg:

> This is the pagan philosophical culture that the early church confronted. It comprised both a contemporary philosophical tradition and a collection of philosophical classics: it dealt with an enormous range of philosophical issues, covering the spectrum from epistemology to politics, and it furnished the technical tools for reasoned discourse. It did not look very much like modern science. (Lindberg, 1986:22)

We must mention yet a fourth element exerting influence on the Catholic tradition, which will be discussed in more detail in the section dealing with the twelfth century. Much of the classical legacy would be lost in the West with the decline of scholarship in the early Middle Ages, but it was preserved in Byzantium and in Islamic cultures after the seventh century. The twelfth-century reintroduction of the full corpus of Aristotle's writings—preserved in Arabic and elaborated upon by Islamic scholars—would enrich the scientific culture of the high Middle Ages. We turn now to look at a few representative writers from the Patristic period.

NATURAL KNOWLEDGE IN THE PATRISTIC ERA

The first six or seven centuries of the Christian Church are called the "Patristic Era," after the great "Fathers of the Church." During this period the church—that had begun as a small Jewish sect—experienced a rapid growth in numbers, first spreading through synagogue communities in the larger cities of the Roman Empire, and then being disseminated throughout the Mediterranean basin as the Christian message attracted gentile converts. For the first three hundred years Christianity was one faith among a welter of competing philosophies and mystery religions and the political religion of the Roman state. Occasionally it was singled out for persecution, the severest episode being that of Diocletian (303–305), during which thousands of Christian martyrs met their deaths in arenas and elsewhere. When Constantine became emperor he issued the Edict of Milan (313) declaring toleration of Christianity, although it became the official religion of the empire only in the reign of Theodosius, in the late fourth century.

In the Patristic Era the Church developed its polity, established its canon of scripture, elaborated its liturgical calendar and patterns of worship, and laid the foundations for important teachings on a wide range of issues. Most significant among these were two key theological problems. The first involved the nature of the relationship between God, the person of Jesus, and the divine spirit working in the Church. This theological argument culminated in the dogmatic definition of God as a Trinity of three persons in one Godhead, worked out at the first ecumenical council at Nicea, convened by Constantine in 325. The second theological problem involved the relationship between Jesus the man and the Christ of faith. Discussion about this lasted into the fifth century, when at the Council of Chalcedon (451) convened by Pope Leo I "the Great" the Church proclaimed that Christ is both fully human and fully divine. The Nicene Creed and the Chalcedonian definition remain touchstones of orthodoxy in all mainstream Christian churches to this day.

Philosophical speculation was not foremost among the concerns of the early Christians, for the gospel was primarily about salvation from a world perceived to be in a "fallen" and morally decrepit state. A young religious community struggling to survive and forge its own identity in the indifferent and often hostile world of the Roman Empire had few resources to spare on the luxury of speculative thought. The Christian New Testament—like the Hebrew scriptures—lacked a systematic interest in natural history. However, in contrast to rival sects such as Gnosticism and Manicheanism, which held dualistic and somewhat pessimistic views of the physical world, the Christianity of the Gospels and Pauline literature remained refreshingly positive about nature. The Genesis affirmation of the world as good is echoed in the Pauline declaration about traces of the divine visible in nature: "ever since the creation of the world his invisible nature, namely his eternal power and deity, has been clearly perceived in the things that have been made" (*Romans* 1:20). This passage would exert a powerful influence on the development of natural theology well into the eighteenth century, serving as one rationale for the study of nature.

From their lack of systematic treatment of natural philosophy and human understanding, the conclusion has sometimes been drawn that the first Christians had no interest in nature. However, as the movement that had begun with fishermen and tradespeople in Palestine began to gather adherents among the educated classes, early apologists felt compelled to articulate the case for the reasonableness of the Christian faith. Justin Martyr (ca. 100 to ca. 165) argued that there are striking parallels between Platonic metaphysical doctrine and Christian belief. He expressed confidence in human reason as the divine gift of "logos," and his philosophical skepticism in fact reflects his appropriation of a respected Aristotelian technique of rational discourse. This was true also of Tertullian (ca. 155 to ca. 230), a North African theologian, who has become unjustly famous for his rhetorical question:

What indeed has Athens to do with Jerusalem? What concord is there between the Academy and the Church?... Away with all attempts to produce a mottled Christianity of Stoic, Platonic, and dialectic composition! We want no curious disputation after possessing Christ Jesus, no inquisition after enjoying the gospel! (Tertullian, 1978:246)

We must read this passage in the context of Tertullian's struggle against Gnosticism and other competing sects. He stood in opposition not to philosophy *per se*, but to philosophy when it led to Gnostic dualism that posited a sharp distinction between a good spiritual realm and the evil material world. Tertullian was careful to defend the possibility of rational knowledge about divine things. Lindberg notes that Tertullian's fideist rhetoric about the resurrection of Christ, "I believe because it is absurd" should not be interpreted as defiant anti-intellectualism. Rather, Tertullian adopted a standard device of Aristotelian logic: the more improbable an event like the resurrection appears to be, the stronger is the probabilistic argument in its favor when it is supported by evidence, such as the eyewitness accounts of the apostles (Lindberg, 1986:26).

Insofar as Patristic literature deals with natural philosophy, it reflects the heritage of Greco-Roman science and the Hebraic-Christian affirmation of the world as God's handiwork. Nature as the de-divinized creation of God was widely regarded as sacramental (particularly in the writings of the Greek Fathers), and as such the natural world was considered a proper subject of investigation. As bishops and theologians, the church fathers were of course only amateur naturalists, but their appreciation of the created world went deeper than merely seeing it as a convenient source for theological metaphors. Basil, Gregory of Nyssa, and Nemesius offered particularly fine objective observations about the plants and animals around them. Botany was studied both through written sources and through the close observation of nature, frequently directed to practical ends such as perfecting the cultivation of fruit trees and staple crops. Patristic zoology, on the other hand, demonstrated the eclectic and unscientific nature of the Fathers, who showed considerable credulity in mixing fact with fancy as they mined the existing literature. Natural history in their hands was not an experimental pursuit. In the sense that they pursued scientific knowledge for its theological and moral use, they demonstrated a closer affinity to Plato than to Aristotle, and the revival of the study of nature as a constructive attempt to understand the animate world in itself would have to await the high medieval recovery of the latter's biological works (Hess, 2002:195–207).

By far the most important patristic writer in the Latin-speaking or Catholic tradition was Saint Augustine of Hippo (354–430). Born in Thagaste in North Africa, Augustine was educated in classical literature and rhetoric, and at the age of seventeen moved to Carthage. Although raised a Catholic by his mother, Monica, Augustine fell under the sway of the popular dualistic Manichean religion, which influenced certain aspects of his mature works. He was in a relationship with a concubine for fifteen years by whom he fathered a son named Adeodatus. In search of more qualified and motivated students, he moved to Rome in 383, but shortly thereafter, he won the post of rhetorician at the imperial court in Milan. Losing faith in Manicheism, he was briefly attracted to skepticism and then to Neoplatonism, which would affect his theology throughout his life. In Milan he came under the influence of Bishop Ambrose and made a dramatic conversion to Christianity (*Confessions*, VIII.xii). Ordained priest and then bishop, he was appointed bishop of Hippo in 396, where he served until his death as Vandal tribes besieged the city in August 430.

Augustine had an ambivalent attitude toward science. On the one hand, he kept it subordinate to his pastoral interests and the goal of leading his flock to salvation, and he cautioned against wasting time on the study of nature that is not useful for the study of divine things (Lindberg, 1986:31). On the other hand, as a rhetorician Augustine made full use of the classical tradition wherever it was relevant, and he regarded familiarity with it as indispensable for a Christian leader who desired credibility in the Roman marketplace of ideas:

If they find a Christian mistaken in a field which they themselves know well and hear him maintaining his foolish opinions about our books, how are they going to believe those books in matters concerning the resurrection of the dead, the hope of eternal life, and the kingdom of heaven, when they think their pages are full of falsehoods and on facts which they themselves have learnt from experience and the light of reason? Reckless and incompetent expounders of Holy Scripture bring untold trouble and sorrow on their wiser brethren when they are caught in one of their mischievous false opinions and are taken to task by those who are not bound by the authority of our sacred books. (Augustine, *De Genesi*, 1.19)

However, even if a Christian ought not to court contemptuous judgments being made upon the faith by pronouncing out of ignorance upon natural philosophical questions, neither ought he to meddle in such questions needlessly. Scientific matters are quite often irrelevant to religious belief:

Nor should we be dismayed if Christians are ignorant about the properties and the number of the basic elements of nature, or about the motion, order, and deviations of the stars, the map of the heavens, the kinds and nature of animals, plants, stones, springs, rivers, and mountains; about the divisions of space and time, about the signs of impending storms, and the myriad other things which these "physicists" have come to understand, or think they have. (Augustine, *Enchiridion*, 3.9)

Augustine notes that even natural philosophers, gifted with superior insight and the leisure to explore the world, still have major lacunae in their

St. Augustine of Hippo (354–430) in his study. Fresco by Sandro Botticelli in the Church of Ognissanti, Florence, 1480. (Harcourt Index)

knowledge. Many mistake what is merely their opinion for certain knowledge, usually on questions not even relevant to the quest for salvation. For the Christian, it should be sufficient "to believe that the cause of all created things, whether in heaven or on earth, whether visible or invisible, is nothing other than the goodness of the Creator, who is the one and the true God" (Augustine, *Enchiridion*, III.9).

However, early Christian scholars did frequently have to field questions involving natural knowledge, as David Lindberg has demonstrated in the case of two examples drawn from Augustine (Lindberg, 2003b:38).

First, like almost all thinkers before the seventeenth century, Augustine believed in the reality of celestial forces and their influence on terrestrial circumstances. Aristotelian thought held that the heavenly spheres are moved by "separated intelligences," which some medieval thinkers would identify with angels. Moreover, since the moon causes the motion of the tides, it is not a huge leap to suppose that the stars also affect mundane affairs. However, Augustine saw that stellar influence on the human mind, as interpreted by astrology, would threaten the Christian doctrines of human freedom, sin, and salvation. He based his opposition to astrology not on theology but on natural science, pointing out that astrologers have never been able to explain the differences one observes in the life fortunes of twins:

Twins are often less like each other than complete strangers; yet, twins are born with practically no interval of time between their births and are conceived in precisely the same moment of a single sexual semination. (Augustine, *The City of God*, 5.1)

Thus, although Augustine's motivation was theological, namely the preservation of human freewill and moral responsibility, he prosecuted his argument against judicial astrology by reference to observable reality.

A second example will demonstrate Augustine's concern to interpret Scripture in consonance with classical physics and cosmology. The puzzle he confronted in his commentary on Genesis was how to account for the waters above the firmament in Genesis 1:6–7: "And God said, 'Let there be a firmament in the midst of the waters, and let it separate the waters from the waters.' And God made the firmament and separated the waters which were under the firmament from the waters which were above the firmament. And it was so." The problem was that in Aristotelian physics, each of the four elements has a natural tendency: earth and water are heavy and have a natural tendency to descend; air and fire being light have a natural tendency to ascend. How is it possible for God to have placed water above the revolving firmament to which the stars are fixed? Augustine's solution was to argue that since water can exist in the form of vapor light enough to be suspended in the clouds, surely it must also be possible for very fine droplets of water to have been suspended above the firmament. He thus developed an interpretation both that was consistent with the literal words of Scripture and that did not violate the principles of Aristotelian physics (Lindberg, 2003b:16–18).

To summarize the relationship between science and religion in the patristic era, Catholic thinkers appear to have chosen a middle path between the enthusiastic pursuit of natural philosophy and the rejection of it; nature was neither to be worshipped nor repudiated. With the changing temper of

Greco-Roman society the philosophical study of nature had already been declining for several centuries, as amply demonstrated by the unsystematic and uncritical character of Pliny's *Natural History*. Philosophical interest in the late empire focused not on the construction of grand interpretive frameworks but on the quest for happiness and spiritual meaning. It is not surprising, therefore, that Christians who participated in intellectual life no more had a monolithic perspective on nature or the appropriateness of studying natural philosophy than did their pagan counterparts. As Lindberg notes, we can speak of centers of gravity along a distribution of positions regarding the relation between scientific and religious concerns. Educated Christians made use of classical natural knowledge for exegesis and for apologetic reasons. In the act of using Greek science, they transmitted it to future generations of scholars, and both theology and natural philosophy were modified in the process. But the transformation was mutual: both Christian doctrine and the philosophical worldview on which it had been built mutually transformed each other. Perhaps most significant was the Church fathers' choice of Platonism: even after the reintroduction of Aristotelian philosophy in the twelfth century, the Platonic idealization of numbers prevailed through the Renaissance to the time of Kepler and Galileo (Lindberg, 2003a:40).

SCIENCE IN THE EARLY MIDDLE AGES: PRESERVING FRAGMENTS

The enormous Roman Empire carried within it the seeds of its own destruction, for as a colossal geopolitical entity it was an enticing target. When in the late fourth century the Huns from central Asia began crowding the barbarian tribes living on the margins of the far-flung provinces, Rome's tottering economic and military infrastructure was no longer sufficient to hold them back. Alaric's Visigoths sacked Rome in 410, the Vandals burned Carthage in 430 and swept into Rome in 455; the Huns ravaged Italy in 452, and the Ostrogoths under Theodoric established a kingdom on the Italian peninsula later in the fifth century. Rome did not fall by one event, but rather collapsed piecemeal, and as her civic and economic life disintegrated, her intellectual culture likewise began to crumble and draw in upon itself.

The Monastic Movement

It was in this context that the institution of monasticism emerged to exert a profound influence on the intellectual development of the Christian West. Two patterns of monastic life had arisen in the fourth century. Eremitical or ascetic hermit monasticism flourished in Egypt, particularly under the charismatic leadership of St. Anthony (251–356). Cenobitical monasticism was developed by Pachomius (ca. 292–348), who had originally lived as a hermit, but who sought to control the excesses of solitary spirituality by promoting a more communal lifestyle. Following Pachomius' death the movement spread to Syria and then around the Mediterranean, where it was promoted in the West by Cassiodorus and John Cassian. The most important western figure in the monastic movement is Benedict of Nursia (ca. AD 480 to 547), who founded twelve monasteries during his lifetime, including the Abbey of *Monte Cassino* in the mountains of southern Italy (Zimmerman, 1980).

In the chaotic social conditions of the early Middle Ages, the Benedictine Order introduced a measure of restraint that assisted in salvaging a partial version of the classical tradition. The *Rule of Saint Benedict* served as both a constitution for the order and a spiritual guide for the monks, and it became one of the most influential documents in Western Civilization. Benedict envisioned a monk's waking life as being divided equally between prayer, work, and study. In his *Rule*, he noted that "idleness is the enemy of the soul; and therefore the brethren ought to be employed in manual labor at certain times, at others, in devout reading" (*Rule*, ch. XLVIII.). This positive valuation of manual labor in the Benedictine order was important for science. In contrast to the classical tradition in which the scholar was exempt from labor, monasticism embraced manual labor, whether it was viewed as a penance or as the practical equivalent of contemplative prayer. Physical work was conducive to the investigation of nature and the venturing of natural philosophical hypotheses, as exemplified by Roger Bacon and Dietrich of Freiburg, and in the nineteenth century by the Augustinian friar Gregor Mendel. Moreover, the monastic approval of technology—in contrast to the Old Testament disapproval of it—led to labor saving devices and the positive appraisal of technology we find in the twelfth century (Kaiser, 1997:81–83).

Two features of monasticism stand out for their impact on medieval science. First, the monastic way was primarily about fostering spiritual conversion of the individual and the community. Genuine conversion is not undisciplined, but is rooted in tradition and learning, and the monks needed the books of their Christian and classical heritage. For this reason, monasticism early on became involved in copying the standard works of theology, history, literature, and natural philosophy. As they worked in their scriptoria laboriously copying and illuminating manuscripts, early medieval monks became the primary force for conserving whatever fragments of liberal studies could be rescued from the collapse of Roman civilization (LeClerq, 1961:112–150). Second because monks sought peace and solitude, their monasteries were often situated in wild and uncultivated places. As they labored to improve the physical situation of their monastic lands, they also improved the situation of the region. This was especially true of the Cistercians (founded in 1098 by Robert of Molesme), who became one of the major vectors of diffusion of technology throughout medieval Europe. It has been argued that Bernard of Clairvaux's (1090–1153) Cistercian emphasis of action in the world for the restoration of the spiritual Jerusalem translated into social planning for the betterment of the human lot, and that this exercised an important influence on the Catholic imagination in the High Middle Ages (Stock, 1975:218–268).

Bestiaries and the Tradition of Natural History

Sharing in the precipitous decline of Western intellectual culture in the period following the collapse of Roman civilization, accurate knowledge about the animate world barely held its own for much of the Early Middle Ages. Although works from this period record considerable familiarity with local plants and their medicinal uses, no significant developments were made in the direction of a systematic understanding of natural history. Fragments of knowledge from the classical tradition were preserved and transmitted in the medieval encyclopedias, such as Isidore of Seville's *Etymologies* and *De proprietatibus rerum* by the thirteenth-century Englishman Bartholomeus Anglicus. Ultimately all such works descended from the anonymous Greek *Physiologus* (ca. AD 200), and their construction is reflective more of literary tradition than of empirical observation, which only infrequently added to excerpts from classic texts.

The knowledge displayed in these works was clearly oriented to didactic and nonscientific purposes. For example, a central theological belief governing the interpretation of natural history was the doctrine that the Fall from grace of Adam and Eve had effected a dramatic physical transformation in nature, including the initiation of carnivorous behavior among animals. Scientific treatment of animals could not simply recount their ecological circumstances or life histories, but had also to pay attention to the natural symbolism of spiritual truths, such as the role of the serpent in the Garden of Eden or of the whale in the story of Jonah, and moral qualities such as the fox's cunning or the dog's fidelity.

The ninth and tenth centuries saw the last significant invasions of Europe by forces from without. The Saracens from the south, the Vikings from the north, and the Magyars from the east harassed Christendom for several generations, until they were repelled, assimilated, or otherwise accommodated by the dominant culture. Europe now entered a period of growing material prosperity, marked by technological improvements that led to increased food production, steady population growth, migration to urban centers, political stabilization, and the foundation of universities.

From the meager and derivative traditions available in the previous millennium, science and the Catholic tradition embarked upon a period of fruitful coexistence.

THE HIGH MIDDLE AGES: THE REDISCOVERY OF ARISTOTLE AND SCHOLASTIC NATURAL PHILOSOPHY

The twelfth and thirteenth centuries hold two stories of great significance for the relationship between science and the Catholic tradition. The first story is the rediscovery, translation, assimilation, and criticism of a much fuller version of the classical tradition than had been available to patristic scholars. As contacts increased between the Latin West and the Byzantine and Islamic worlds, so did the awareness on the part of European scholars of the intellectual heritage that had been lost in the collapse of Roman civilization in the West. The translation of scientific and other texts from Greek and Arabic into Latin by scholars working in Italy and Spain offered renewed access to the full classical legacy. These texts included the complete corpus of Aristotle's writings, as well as other ancient works and more recent Byzantine and Islamic commentaries on them. Latin scholars were not merely passive recipients of this material, but augmented it in terms of both content and methodology (Colish, 1997:319; Grant, 1996:18–32).

The second story is the foundation of medieval universities and their rapid growth in number and organization. The University of Bologna was founded in 1158, Paris and Oxford shortly thereafter, and two dozen more had sprung up in Europe by the end of the thirteenth century. The university curriculum developed from the undergraduate training in the *trivium* or "threefold way" of grammar, logic, and rhetoric (Leff 1992:307–336) and the *quadrivium* or "fourfold way" of arithmetic, astronomy, geometry, and music (North 1992:337–359), which were legacies of the meager classical tradition that antedated the universities, and grew to include graduate degrees in law, medicine, and theology. The approach to learning taught in universities until the seventeenth century, known as "scholasticism," was based upon a process of dialectical reasoning on disputed questions that were explored by marshalling authorities on either side. One feature of scholasticism was the writing of *summae*, or compendia of wide bodies of knowledge, the most famous of these being the *Summa Theologiae* ("sum of all theology") of Saint Thomas Aquinas.

Together, the establishment of universities as powerful new institutions of learning and the recovery of the classical tradition exercised a profound impact on the relationship between science and theology. Of course, the scholastic embrace of Aristotelianism was not without its challenges. As we have seen, Platonism and Neoplatonism had appealed to Augustine and other patristic writers as being essentially more compatible with Christian doctrine than the principal alternative. Aristotelian thought carried serious impediments, including the idea of the eternity of the world, which ruled out creation in time. Moreover, the idea of a God who was uninvolved with creation was not compatible with personal theism, and Aristotle's theory of the soul left no room for life after death. However, when purged of its most objectionable elements, Aristotelianism—with its syllogistic logic and its empirical approach to natural philosophy—served as a powerful tool for understanding the world.

An important early exponent of Aristotelianism was the Dominican friar Robert Grosseteste (ca. 1175–1253), who integrated new scientific insights into his Christian Neoplatonism (Colish, 1997:319–321). He participated in translating Aristotle's science and ethics and maintained his interest in the new curriculum while Chancellor of Oxford University and later Bishop of Lincoln. In contrast to later scholastics, Grosseteste was more interested in Aristotle's rationalist approach to science than he was in developing an integrated worldview. He adopted Aristotle's assumption that four causes—material, formal, efficient, and final—are responsible for the operation of all things, and reasoned that generalizations (induced from the observation of particulars) lead to conclusions that are then tested by means of deductive logic. But a significant innovation on Aristotle's demonstrative syllogism was Grosseteste's pursuit of controlled experiments that can provide mathematically quantifiable and measurable data. It was this mathematized view of the world that four centuries later would make possible Galileo's measurement of the rate of acceleration of balls rolling down inclined planes. Grosseteste also combined a fascination with the working of nature with interest in its technological applications. His experimental work with angles of refraction in glass spheres, for example, led to the invention in his own century of corrective eyeglasses, and ultimately to the telescope that would revise our understanding of the heavens.

Better known even than Grosseteste for his embrace of Aristotelianism as a productive methodology was his pupil Roger Bacon (ca. 1214–1294), an English natural philosopher committed to empirical methods. An independent arts master who entered the Franciscan order later in life, Bacon found his activities restricted by the Franciscans, and he no longer held a teaching post after 1260. Among the most important dimensions of Bacon's work was his composition of three treatises (written in the 1260s at the request of Pope Clement IV) for the reform of learning in Christendom. The challenge was to chart a middle path between uncritically embracing new material that could be employed to serve only rational and secular ends, and repudiating the legacy of this new learning in fear of the potential dangers it posed to Christian doctrine (Lindberg, 2003b:21–29). In the

Opus Maius (*Larger Work*), Bacon articulated the idea that philosophy is not an autonomous enterprise, but exists as a handmaiden to serve the Church and its theology. In Bacon's liberal perspective, however, there was very little in the new learning that did not legitimately perform this function.

An equally appreciative (if more cautious) student of Aristotle was Albert of Lauingen in Saxony, known as Albertus Magnus (Albert "the Great," ca. 1200–1280). He was a valued administrator for both the Dominican Order and the Church, being made provincial of Teutonia and then Bishop of Regensburg. Discharging his duties conscientiously, Albert was at heart always a scholar. Although theology was his principal interest, he regarded natural philosophy as a complementary way of examining the created works of the First Cause (Synan, 1980:12). In an atmosphere that all too readily made Aristotle a canon of truth, Albert quite candidly asserted his own independence:

Perhaps some will say that we have not understood Aristotle and that on this account we have not agreed with what he said or that (from their certain knowledge) we contradict him in point of truth on some matter. To him we say that whoever believes Aristotle was a god ought to believe that he never erred; if, however, one believes him to be but a man, then without doubt he could err just as we can too. (Synan, 1980:11)

The Order sent Albert to Cologne in 1248 (accompanied by Thomas Aquinas) to establish a *studium generale* (general order of study) for Dominican students in Germany. Quite boldly, Albert lectured on Aristotle's *Physics* in the stadium, convinced that a solid grounding in both philosophy and science were indispensable to theological studies. This was the beginning of his ambitious project of laying out in a systematic way the whole of human learning, including not only the contents of the *trivium* and *quadrivium*, but of the natural sciences, economics, ethics, politics, and metaphysics (Weisheipl, 1980:30).

The most famous of the scholastic philosopher-theologians was Thomas Aquinas (1225–1274), an Italian Dominican whose *Summa Theologiae* and *Summa Contra Gentiles* would become canonical texts for Catholicism after the Council of Trent. Perhaps more than anyone else Thomas helped to convert the building blocks of Aristotelian logic into the scholastic method. Theology and natural philosophy (now science) were carefully distinguished from each other, and taught in separate university faculties (Grant, 2001:185–186). A careful scholar, Aquinas emphasized the provisional nature of the interpretation of natural philosophy, prudently noting the different starting points of philosophy and theology:

Human philosophy considers creatures as they are in themselves: hence we find different divisions of philosophy according to the different classes of things. But

Albertus Magnus (St. Albert the Great, 1200–1280). Fresco by Tommaso da Modena, Treviso, 1352. (Image courtesy History of Science Collections, University of Oklahoma Libraries; copyright the Board of Regents of the University of Oklahoma)

Christian faith considers them, not in themselves, but inasmuch as they represent the majesty of God, and in one way or another are directed to God, as it is said: *Of the glory of the Lord his work is full: hath not the Lord made his saints to tell of his wonders?* (Ecclus xlii, 16, 17.) Therefore the philosopher and the faithful Christian (*fidelis*) consider different points about creatures: the philosopher considers what attaches to them in their proper nature: the faithful Christian considers about creatures only what attaches to them in their relation to God, as that they are created by God, subject to God, and the like. Hence it is not to be put down as an imperfection in the doctrine of faith, if it passes unnoticed many properties of things, as the configuration of the heavens, or the laws of motion. And again such points as are considered by philosopher and faithful Christian alike, are treated on different principles: for the philosopher takes his stand on the proper and immediate causes of things; but the faithful Christian argues from the First Cause, showing that so the matter is divinely revealed, or that this makes for the glory of God, or that God's power is infinite. Hence this speculation of the faithful Christian ought to be called the highest wisdom, as always regarding the highest cause, according to the text: *This is your wisdom and understanding before the nations* (Deut. iv, 6). (*Summa Contra Gentiles*, II.4)

Aquinas noted that human philosophy is subordinate to the higher wisdom that is revealed truth, and that philosophy and theology proceed in a different fashion. Philosophy begins from the premise of the created order, and from the study of created things it is led on to metaphysics and the knowledge of God. Theology, in contrast, is more reflective of the divine mode of knowing, studying as it does first God and only afterwards the creation.

Efforts to control the more threatening aspects of Aristotelian natural philosophy led to a growing rift between the radical arts masters such as Siger of Brabant and Boethius of Dacia, and the more traditional and conservative theologians such as the Franciscan Bonaventure. The former regarded Aristotle's method as the key to a proper understanding of the universe, and regarded philosophy as independent of and perhaps superior to theology (Grant, 1986:54). Attempts were made to ban or expurgate Aristotle's books early in the thirteenth century, but by 1255, they constituted the core of the arts curriculum in a number of universities. The traditionally minded neo-Augustinian theologians took the drastic step of appealing to Stephen Tempier, Bishop of Paris to review the matter. Tempier responded in 1277 with an extensive condemnation of 219 propositions drawn from a variety of philosophical sources, including the works of Thomas Aquinas:

Lest, therefore, by this unguarded speech simple people be led into error, we, having taken counsel with the doctors of Sacred Scripture and other prudent men, strictly forbid these and like things and totally condemn them. (Grant, 1974:44–45)

Of particular concern to opponents of scholastic natural philosophy was the danger posed to the Christian worldview by the tradition of Averroes (Ibn Rushd, 1126–1198), the great twelfth-century Arabic philosopher from Cordoba, in Spain. In a fashion parallel to scholasticism, Averroes attempted to integrate Aristotelian thought with Islam. Traditionalists would find particularly threatening his upholding of Aristotle's eternity of the world and his division of the soul into divine and human parts.

Edward Grant has noted that the impact of the Condemnations on science in the Catholic tradition was ambiguous. The articles directed against

S. THOMÆ AQVINATIS

In Duodecim Metaphysicorum libros
ARISTOTELIS,

COMMENTARIA CELEBERRIMA.

CVM DVPLICI TEXTVS TRALATIONE
antiqua & noua Bessarionis Cardinalis:

Fr. BARTHOLOMAEI SPINAE ORDINIS PRAEDICATORVM
Disertissimæ defensiones locorum ab Ant. Andrea impugnatorum.

(um Argumentis singularum Lectionum, & Indice copioso:

QVAE OMNIA IAM DILIGENTIVS RECOGNITA,
à multis quoque mendis & erroribus fuisse purgata Lector gaudebit.

VENETIIS, Apud Hæredem Hieronymi Scoti. 1588.

St. Thomas Aquinas. (Image courtesy History of Science Collections, University of Oklahoma Libraries; copyright the Board of Regents of the University of Oklahoma)

Thomas Aquinas were nullified in 1325, but the rest of the condemnation remained in effect throughout the fourteenth century. Prima facie, this might seem like a clear case of ecclesiastical overreach putting the brakes on scientific speculation. However, on a closer view things appear differently. The articles were written to condemn illegitimate limitations placed on the power of God by the arts masters, who had chastised anyone who dared argue with Aristotle. Article 147, for example, condemned the proposition "that the absolutely impossible cannot be done by God or another agent," where "absolutely impossible" refers to things contrary to the Aristotelian worldview (Grant, 1986:54). Grant has argued that limiting the pretended authority of Aristotle in fact freed natural philosophers to entertain imaginative thought experiments (*secundum imaginationem*) in unorthodox or even anti-Aristotelian directions. Appeals were made to the *potentia Dei absoluta*, God's absolute power, as distinguished from the *potentia Dei ordinata*, the manner in which God actually disposes of divine power. Might God (contrary to Aristotle) have created a plurality of worlds? Could God move a world in a rectilinear direction in space such that in its place would be left a vacuum? And if God could create a vacuum between worlds, could God create a vacuum on earth? Might a moving earth revolve around a stationary sun?

LATER SCHOLASTICISM: EXPLORING NEW AVENUES

Lindberg has cautioned us that that "the extreme claims about the medieval anticipation of early modern developments in science are simply false" (Lindberg, 1992:360). However, it is fair to say that by opening up a space for the contemplation of provocative questions, the Condemnations of 1277 mark an important point in the evolving relationship between Catholicism and the sciences. Stephen Gaukroger notes, for example, that Archbishop Tempier's repudiation of the Aristotelian assertion that God could not have created another world virtually invited discussion of this very possibility. The great polymath Nicole Oresme (1323–1382), a philosopher and mathematician later appointed Bishop of Lisieux, argued that although God has in fact created only the one world we know, he could in fact have created several. His consideration of the physics of multiple worlds was rather insightful, noting that all particles of matter in the earth constitute one mass, numerically speaking:

And if some part of the earth in the other world were in this world, it would tend toward the centre of this world and become united with the mass, and conversely. But it does not follow that the portions of earth or of the heavy bodies of the other world, were it to exist, would tend to the centre of this world because in their world they would form a single mass possessed of a single place and would be arranged in up and down order, as we have indicated, just like the mass of heavy bodies in this world. (Oresme, 1968:174, cited in Gaukroger, 2006:74–75)

Oresme has imagined his way right past the Aristotelian doctrine of there being only one possible "natural place" for earthy objects, one center of attraction, and has moved toward a physics of multiple centers of gravitation such as Newton would articulate four centuries later. Oresme also anticipated Copernicus by considering the question of whether it is the earth or the sun that moves. He noted that since motion is relative, we simply cannot establish empirically whether it is the heavens or the earth that rotate daily. Considering the scriptural texts in question (see Chapter 2), he anticipated Galileo's argument that these passages are not to be read literally. Oresme pointed out that it is simpler for the earth to revolve on its axis than for the entire heavens with its massive bodies to revolve daily, although ultimately he concluded in favor of the geocentric model.

An important characteristic of later medieval scholasticism was its reconciliation of Aristotelian natural philosophy with theology couched in Neo-Platonic terminology. Alexander Murray has spoken of the twofold process of integrating Greek thought into western consciousness: "it was Christianity's otherworldly self that digested Plato and his tradition; and its this-worldly self that digested Aristotle" (Murray, 1992:18). Theodoric of Freiberg (1250–1310), for instance, was Neoplatonic in his theology but Aristotelian in his natural philosophy. Possibly the first person to offer a correct explanation of rainbows, he simulated water droplets by using water-filled spheres, and measured the refraction of light as it entered and exited the spheres (Wallace, 1959:183–225). Oresme likewise, while remaining firmly planted within an Aristotelian natural philosophical framework, used a number of Platonic or Platonically inspired objections to criticize specific details of the Aristotelian worldview (Gaukroger, 2006:88).

A century later, Nicolas of Cusa (1401–1464), a German philosopher, mathematician, and astronomer later created cardinal, developed interesting arguments in *De Docta Ignorantia* (*Of Learned Ignorance*). Against Aristotelian-Ptolemaic astronomy, he claimed that no perfect circle could exist in the universe, anticipating Kepler's insight about elliptical planetary orbits. Cusa contributed the concept of the infinitesimal to mathematics, and corrected myopia with concave lenses. The unity of his vision is illustrated by his applying the famous saying that God is "a sphere of which the centre is everywhere and the circumference nowhere" (originally from a twelfth-century pseudo-Hermetic treatise) to the universe as a whole. For Nicolas of Cusa, the universe is a reflection of God, as it would later be for Giordano Bruno, "for whom the innumerable worlds are all divine centers of the unbounded universe" (Yates, 1964:247).

CONCLUSION: FROM LATE SCHOLASTICISM INTO EARLY MODERNITY

For much of the first millennium of the Catholic Christian tradition, culture and learning stood at a very low ebb. At first preoccupied with establishing itself in the hostile environment of the Roman Empire, after 313 and the Edict of Milan the Church was absorbed with articulating its theology and its polity. When the Roman Empire collapsed with the barbarian invasions of the fifth century, the Church stepped into the breach to help hold together what remained of the infrastructure and to preserve what remained of the fragments of civilization, and "science" consisted mainly of copying, commenting upon, and transmitting the "thin" classical tradition. So little was known about how the world actually works that there were few points of friction between natural philosophy and Christian theology articulated through Platonic metaphysics.

With high medieval prosperity fostering the growth of urban populations, the conditions were propitious for founding universities, coinciding with the recovery and translation of the full Aristotelian corpus. Scholastic university culture held as its ideal the *unitas scientiae*, or "unity of knowledge," approaching the universe as a coherent knowable unity. Aristotelianism quickly gained ascendancy and served as interpretive matrix for both philosophical and theological questions (Murdoch, 1975:308). However, as the firm grip of Aristotelianism became loosened in the fourteenth century, avenues were opened to speculation in directions that would eventually lead to the ideas of Copernicus and Kepler, Bruno, and Galileo. The "*unitas scientiae*" would endure as an ideal for another four centuries (although increasingly fragmented) in early modernity, until with the seventeenth-century professionalization of the sciences it would become largely supplanted by the compelling lure of more practical objectives. But for Catholic thinkers (as no doubt for others) the unity of knowledge is an ideal, an asymptote to be striven for, even in our own age of disciplinary compartmentalization.

Chapter 2

From Cosmos to Unbounded Universe: Physical Sciences from Trent to Vatican I

INTRODUCTION

In 1543, as he lay on his deathbed in Frombork, Poland, Nikolaus Copernicus was handed a copy of his newly published *De revolutionibus orbium coelestium*. Two years later, in December 1545, a long-delayed ecumenical council was convened by Pope Paul III and called into session in the cathedral of Trent, in northern Italy. These events symbolize two great revolutions that would affect the relationship between Catholicism and the sciences over the next four centuries. The council of Trent constituted in part the formal response of the Roman Catholic Church to the challenge posed by Martin Luther in 1517. Expanded with the reforms of John Calvin in Geneva, the radical reformers in Germany, Thomas Cranmer in England, and John Knox in Scotland, the Reformation would sunder the unity of medieval Christendom. Copernicus's book initiated a process that led to a great revision of the Western view of the cosmos. He unwittingly ushered in sweeping changes in scientific methodology and understanding, in a movement that would come to be known as the Scientific Revolution.

Discussing the relationship between science and religion during a time of profound transformation in the assumptions underlying and the methods for studying the natural world involves us in a potential minefield. The Scientific Revolution is notoriously difficult to define, and has even been challenged as a meaningful concept. The relationships between religion and science during this pivotal period—in both theory and practice—were many and intricate. It is crucial that we note the qualitative differences between the hypothetico-deductive enterprise we know as empirical science in the twenty-first century on the one hand, and on the other the predominantly nonempirical natural philosophy of the West prior to 1550. If "Scientific Revolution" is used to connote a prepackaged and inevitable triumph of modern science as we understand it, the critics are right and it is a misleading concept. However, it is a useful model if by it we understand the series of related events and intellectual trends—spread over three centuries—that have radically changed the way humans understand the world (Osler, 2000:3–22).

Every era has had its "science," its way of understanding the world through physics, astronomy, and the studies of the earth and of life. Christian theologies understandably have been intricately bound up with the cosmos as they understood it, but of course an imprudent entanglement carries an implicit hazard: too close a dependence of theology upon any particular worldview entails the risk of leaving that theology adrift when the worldview changes radically. Conversely theology can scarcely be intelligible if it remains wholly apophatic about the world (Wildiers, 1982:158–160). The medieval cosmos eventually gave way to the seventeenth-century universe, and this in turn ceded to various modern and now postmodern syntheses. Every disintegration of a worldview gives rise to a corresponding reintegration of the remnants into a new world-view, and at each step the Catholic tradition has responded—albeit with varying degrees of caution—by gradually appropriating the dominant ideas of the relevant sciences.

THE REFORMATIONS OF THE SIXTEENTH CENTURY

The Impact of the Protestant Reformation

On October 31, 1517, an earnest young Augustinian canon and scripture scholar, Martin Luther, nailed his *Ninety-Five Theses* to the door of the Castle Church in Wittenberg. The immediate provocation was the preaching of indulgences by an itinerant Dominican friar, Johann Tetzel, which offended Luther both as a proud German and as a theologian concerned about the abuses involved with this practice. An indulgence was originally an offer of remission of the temporal punishment for a sin; later it was extended to apply to purgatorial punishment as well. In Luther's day an indulgence was offered in exchange for a monetary payment to the Church (Pelikan, 1971:134–137). Specifically, what Luther called for in his *Theses* was an open and honest discussion of the nature of penance, the authority of the pope, and the usefulness of indulgences for salvation. The more general provocation was the alarming climate of ecclesiastical corruption and theological ambiguity that had crept upon the Church since the Late Middle Ages (Ozment, 1980:204–222). Luther wrote as a faithful

Catholic theologian, and hoped that his sincere call for honest and open discussion of the need for both theological and disciplinary reform within the Church would be heeded.

Regrettably, reform that was both timely and thorough enough to preserve the unity of Christendom was never achieved. For geopolitical reasons beyond the control of the principal actors, by the time the Church acquiesced to discussion of reform it was too little and too late. Luther had issued his call for discussion about indulgences in 1517. Not surprisingly, Rome, instead of agreeing, had initiated an inquisition into Luther's activities in 1518, but became distracted by the death of Emperor Maximillian in 1519. In public debate Luther was forced to declare his differences with the Church, and expanded his critique with three far-reaching treatises: *On the Freedom of a Christian* (dealing with the theology of justification), *The Babylonian Captivity of the Church* (a broad critique of the sacramental system of the Church), and *Address to the Nobility of the German Nation* (a statement of the causes of growing social discontent in Germany). The appearance of these tracts had pushed Rome too far, and Pope Leo X issued the papal bull *Exsurge Domine* (June 15, 1520) that forbade the printing, selling or reading of Luther's works. Luther responded by publicly burning the bull and the books of canon law on December 10, and Leo excommunicated Luther on January 3, 1921.

Maximilian's successor, the young Holy Roman Emperor, Charles V (1500–1558), convened a general assembly of the estates of the empire, called the Diet of Worms, in the spring of 1525, in order to address the crisis of Luther and his proposed reformation. No quarter was shown to the heretic monk, however, and Charles issued the Edict of Worms, effectively putting a price on the reformer's head. On his return journey from the diet, Luther was kidnapped for his own protection and confined by Prince Frederick the Wise in the Wartburg Castle. There Luther began his translation of the Bible into German, and during the next five years he became married, composed vernacular baptismal and wedding services, and wrote catechisms and other works for the Protestant church. The Reformation continued to attract converts and gathered strength throughout the German estates. At the imperial Diet of Speyer in 1529, the followers of Luther and the other reformers were first called "Protestants," and in 1531 an alliance of Protestant princes formed the Smalkaldic League. Thus, by the time the Council of Trent was convened in 1545—ironically only a year before Luther's death—German Protestantism was a well-established fact, and hopes of reuniting a Christendom purified in doctrine and practice had all but died. John Calvin's reform of the Church in Geneva and Henry VIII's revolt against Rome ensured that Reformation was secure, and thus the Council of Trent was forced, in part, to define itself in reaction to Protestantism.

The Catholic Response

Historians have long debated whether we should speak of Counter-Reformation, Catholic Reformation, or Catholic Renewal or Revival (Iserloh, 1980:431–432, 1946). There is sufficient evidence to support both interpretations: that the Catholic Reformation was essentially a tardy and begrudging effort to stem the hemorrhage to Protestantism, and that it was an earnest and sincere endeavor to work out internal reforms that had begun before Luther (Gleason, 1981:3–5). A balanced answer would no doubt incorporate elements of both readings (Hsia, 2005:1–9; Mullett, 1999:ix–x, 1–28;). However, no matter how we assess the larger movement of Catholic renewal, it is undeniable that the Council of Trent (1545–1563) was a movement to fortify the metaphorical walls of Rome against Protestantism.

When Charles V recognized the seriousness of Luther's revolt, he agreed with the German princes and called for a general council. Pope Clement VII (1478–1534) was ardently opposed to the idea, echoing the consistent papal opposition to the Council of Constance's (1414–1418) declaration of the supremacy of councils over popes. By the mid-1530s, however, it became clear to Pope Paul III (1534–49) that the Reformation was not a local and temporary phenomenon, and that if he wanted to save any of the German principalities or city-states for Rome, he must acquiesce to a general discussion. Over the opposition of the cardinals, he called for a council to meet in Trent, which as a free city of the Holy Roman Empire would permit greater neutrality. The nineteenth Ecumenical Council of the church met there in three sessions between 1545 and 1563, with most of the creative work coming in the final session. Achievements of the council included the theological definition of the Catholic doctrine of salvation, a formal statement on the canon of the Bible, and clarification and reinforcement of the sacramental system that Luther had challenged. Trent also standardized the order and form of the Eucharistic celebration, and the Tridentine form of the Mass endured until it was changed as a consequence of the Second Vatican Council, first convened in 1962.

In the context of our discussion of the relationship between Catholicism and science, three aspects of the Catholic Reformation stand out as being of signal importance. First is the promulgation of the decree on Holy Scriptures. The Protestant principle of *sola scriptura*—"by scripture alone"—was central to the thought of all the reformers. Luther, Calvin, and Ulrich Zwingli all had emphasized the primacy of personal interpretation of scripture, the only authoritative source of Christian doctrine: they regarded the Bible as self-explanatory, requiring no clerical hierarchy or ecclesiastical tradition to interpret it. Regrettably, human fallibility suggests that there would be as many interpretations as there are interpreters, and the divisive tendencies inherent in "sola scriptura" became painfully apparent at the Reformers' Marburg Colloquy of 1529 about the meaning of the Eucharistic presence. At Trent the Catholic Church had strengthened its long-held conviction that authority over the interpretation of scripture rests with the magisterium of the Church, that is the teaching authority of the bishops in communion with the pope, the successor of St. Peter:

Furthermore, to check unbridled spirits, it [the council] decrees that no one relying on his own judgment shall, in matters of faith and morals pertaining to the edification of Christian doctrine, distorting the Holy Scriptures in accordance with his own conceptions, presume to interpret them contrary to that sense which holy mother Church, to whom it belongs to judge of their true sense and interpretation, has held and holds, or even contrary to the unanimous teaching of the Fathers, even though such interpretations should never at any time be published. (Schroeder, 1978:18–19)

The far-reaching significance of this canon cannot be overemphasized, but it must be placed in historical context. Tight control over how the faithful may read the Bible may seem incomprehensible to postmoderns, but even the Reformers recognized how an idiosyncratic reading of scripture could lead to extremes (as in the case of John of Leiden, who briefly presided over a theocracy of his own devising in Münster, beheading a disobedient wife (Haude, 2000b:16). Rome was in the process of losing cities and even principalities to the Reformation in the blink of an eye, and it was not about to open the door to personal interpretation of scripture by everyone else who still remained within the fold. As we shall see, this issue would become of great significance sixty years later in the case of Galileo.

A second consequence of the Counter-Reformation significant to the study of Catholicism and science was the centralization of the Inquisition and the establishment of the *Index of Prohibited Books*. The Church had exercised its authority over internal threats since its earliest days, and inquisitions against both individual and collective heresies had been carried out since the twelfth century. Pope Paul III established the Congregation of the Holy Office in 1542 to serve in a supervisory role over local inquisitions. Now more than ever the Church perceived it as crucial to defend the integrity of doctrine against errors and heretical teachings. One of the most famous victims of this formalized Inquisition of the Counter-Reformation was Giordano Bruno (1548–1600), although contrary to popular opinion he was burned at the stake in Rome not for his scientific ideas but for theological heresy, namely, denying the divinity of Christ.

Another issue facing Catholicism was the flood of books and pamphlets inundating Europe, as the printing press had made it possible for virtually anyone to publish his opinions. In the face of Protestantism the Church and its academic affiliates felt the need to exercise more control over the sale and possession of threatening materials, and in 1544 the first list of prohibited materials was published by the theological faculty of the University of Paris (Artigas, 2006:9). The Council of Trent further systematized the process of censorship, and in 1564 Pope Paul IV promulgated the *Index Librorum Prohibitorum* ("List of Prohibited Books"), which included works regarded as potentially harmful to the faithful for reasons of moral or doctrinal error. Within a few decades, at least one astronomer would feel the power of both the Inquisition and the Index.

A third outcome of the Catholic Reformation that would be important to continued Catholic involvement in science was the foundation of the Jesuit order. The spiritual renewal surrounding the era of reformations prompted a proliferation of new religious congregations responding to the needs of early modernity. Among the most significant of these was the Society of Jesus, founded in 1543 by Ignatius of Loyola (1491–1556), a noble Spaniard of Basque lineage, and six of his friends. They called themselves the "Company of Jesus" (or "companions of Jesus") and dedicated themselves to the special service of the pope and in particular to missionary work. Headed by a general, not an abbot, they were an active and well-organized order, perfectly suited for the evangelization both of Europe sundered by the Reformation, and of the peoples around the globe discovered in the course of the European reconnaissance (O'Malley, 1993:52–61). The Jesuits were superb teachers and educational administrators, and by the time of Ignatius's death they had already established seventy-four colleges on three different continents, a number that rose to 245 by the turn of the century (Artigas, 2003:5). Jesuits figured prominently among scientists in the late Renaissance, including the chief architect of the Gregorian calendar reform, German-born astronomer and mathematician Christopher Clavius (1538–1612).

THE UNMAKING OF THE MEDIEVAL COSMOS: COPERNICUS REVISES THE HEAVENS

In the popular understanding, what most paradigmatically represents the Scientific Revolution of the sixteenth and seventeenth centuries is the great transformation in astronomy from the Ptolemaic cosmos to the Newtonian universe. Early modern Europe experienced many shifts in perspective, ranging from physiology and natural history to chemistry, geology and other sciences. But the revolution in cosmology was the most visible and dramatic, and it dealt with the heavens, a subject relevant not only to science but also to religion. We must bear in mind that although modernity has sterilized the study of "the heavens" of all but purely astronomical content, late scholastics distinguished between three interrelated concepts: (1) the Empyrean, a theological concept concerning heaven as the abode of God and the saints, (2) the Firmament, a biblical idea about the waters below and the waters above that caused the Noachian flood, and (3) the physical space in which revolved the seven planets, a concept subject to "scientific" or natural philosophical speculation (Randles, 1999:1–8). It is thus a mistake to assume that the Catholic Church should have been uninterested in Galileo's campaign to interpret his telescopic findings. Moreover, the revolution in cosmology did not remain only that; the challenge to the received tradition of medieval astronomy served as a powerful catalyst for what would become revolutions in thinking about every dimension of the world, from Aristotelian physics to Galenic physiology and medicine.

The late medieval discussion of astronomy took a decisive turn with the work of Mikołaj Kopernik, born in Torún, Poland, on February 19, 1473. Known to us as Nikolaus Copernicus, he was raised in the comfortable circumstances of a wealthy burgher family and was educated at the cathedral school (Hess, 2004:1976–1979; Rosen, 1984:55–74). When his father died in 1483, Nikolaus and his younger brother Andreas were taken under the guardianship of his maternal uncle, Canon Lucas Watzenrode, who had been trained in the cosmopolitan humanist atmosphere of Bologna and later was appointed Prince Bishop of the Diocese of Warmia. Copernicus matriculated in the renowned Jagiellonian University of Cracow, which was strong in mathematics and boasted an endowed chair of astronomy. His study of the theories of such luminaries as Ptolemy, Euclid, Sacrobosco, and Regiomontanus was complemented by his own observation in Cracow of the comets of 1491 and 1492, and of four lunar and solar eclipses during the next two years.

Watzenrode sent the Copernicus brothers to Bologna in 1496 to further their education. During his decade in Italy Copernicus continued his observations of the heavens, became well versed in philosophy and classical literature, studied medicine at Padua, and in 1503 took his degree in canon law from the University of Ferrara. Watzenrode had arranged for his election to a benefice in the diocese of Warmia to ensure his nephew's financial independence, so Copernicus returned to Poland to embark upon his duties as a canon of Frombork Cathedral. Although not a priest, for the next forty years he was engaged in ecclesiastical administration and other services to the diocese, and also found time to write an important treatise on coinage, to paint a self-portrait, and conscientiously to practice medicine. Astronomy remained his passion, however, and in 1510 he built a modest observatory in a tower near the cathedral.

Copernicus' rehabilitation of the long-neglected heliocentric hypothesis altered astronomical perspective and led to the transformation of science

Nicholas Copernicus (1473–1543). (Image courtesy History of Science Collections, University of Oklahoma Libraries; copyright the Board of Regents of the University of Oklahoma)

as a whole. The concept of a moving earth was not new: it had been proposed in the third century BCE by Aristarchus of Samos and was discussed by Archimedes and by the Pythagoreans Philolaus and Ecphantus (Lindberg, 1992:97–98). But the classical arguments against heliocentrism—both from common sense and from the lack of observed stellar parallax—had been so overwhelming that the alternative geocentric cosmology prevailed from antiquity through the later Middle Ages. Aristotle envisioned a set of nested concentric spheres bearing the planets, the sun, the moon, and the stars in circular orbits around a stationary earth. This system was enlarged and codified in the second century by the Egyptian astronomer Ptolemy, whose *Almagest* became the basic astronomical text of the scholastic university curriculum, enduring well into the seventeenth century (Costello, 1958:7–11; 102–104). Ptolemy sought both to preserve uniform circular motion of the spheres and to account for the periodic retrograde motion of the planets against the backdrop of the fixed stars. To achieve this, his system invoked a scheme of epicycles and eccentrics that revolved on deferent circles about an equant point that itself revolved around the earth (Lindberg, 1992:98–105).

In scholastic science Ptolemaic astronomy was integrated with Aristotelian physics. This model declared that circular motion was proper to the heavens and rectilinear motion to the earth—hence the earth could not possibly rotate on its axis. In Aristotelian science the four terrestrial elements (earth, water, fire, and air) were arranged appropriately in relationship to one another according to their degree of levity or gravity. The stars and planets (Latin *planetes* means "wandering stars") were composed of yet a fifth essence, a "quintessence." The pre-Copernican cosmology was in turn integrated with theology to form an orderly scholastic synthesis of physics, astronomy, and theology. Each science in this hierarchy of disciplines operated from its own set of principles, and together they governed everything from the nature of matter and planetary motion to the geographic location of heaven and hell. The topography of Dante Alighieri's *Divine Comedy* reflects the structure of the Aristotelian-Ptolemaic cosmos.

However, as we now recognize, neither the cosmos nor our human understanding of it stands still, and by the sixteenth century, Ptolemaic astronomy had begun to encounter difficulties in accurately predicting celestial phenomena. The complex system of eccentrics and epicycles revolving about the equant point "saved the appearances," and for centuries had proven to be reasonably reliable in accounting for retrograde planetary movements without sacrificing uniform circular motion. But the minute inaccuracies of this system—when compounded annually over more than a millennium—had gradually pushed astronomical reckoning off by ten days. This cumulative error posed serious calendrical problems, including the difficulty of correctly calculating the date of Easter, the central Christian feast on which much of the church year was based. As religious observance (along with living a morally upright life) was believed to be important to the salvation of believers, astronomy for this reason was quite relevant to religious practice.

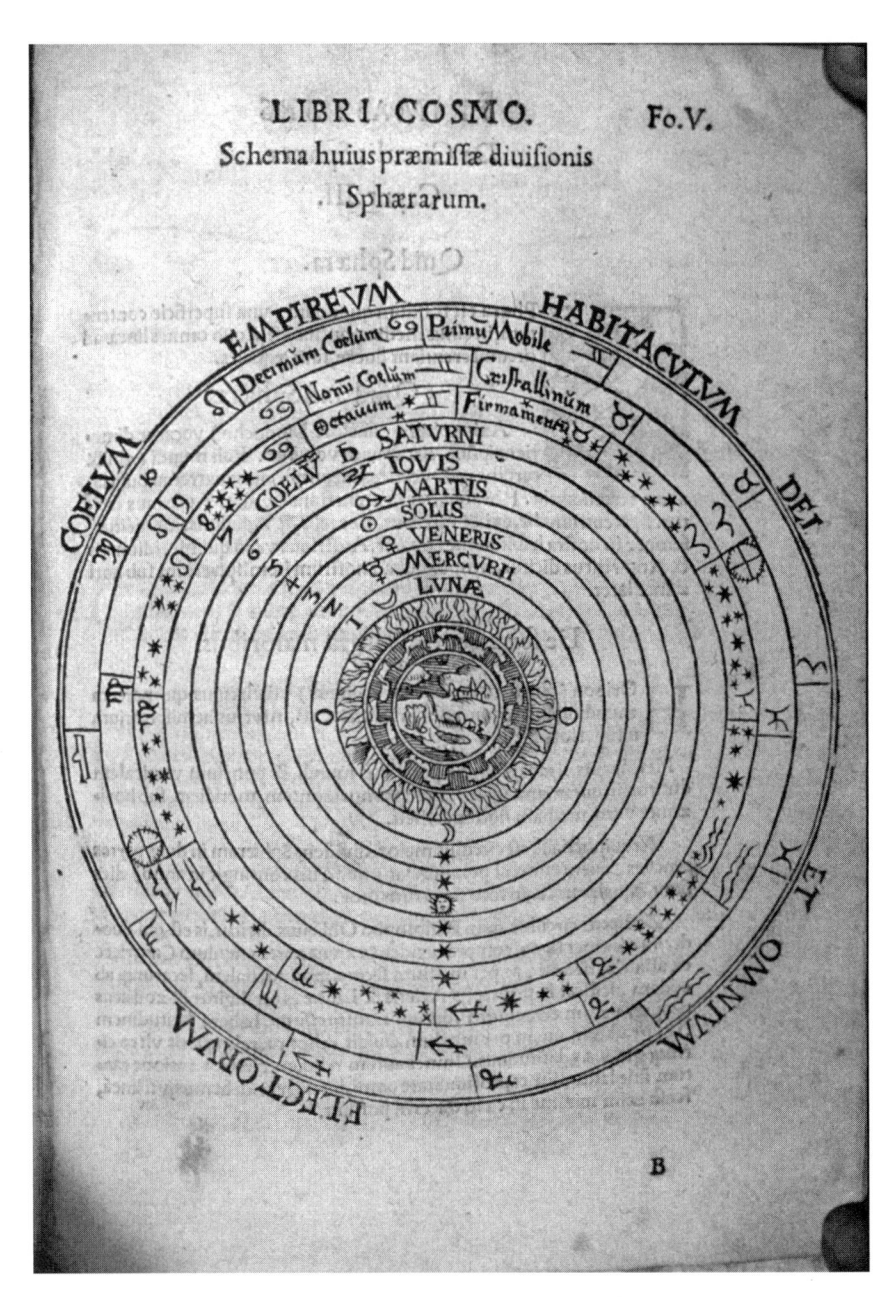

One rendering of a Ptolemaic model of the cosmos, from Peter Apian's *Cosmographica* (Antwerp, 1539). (Image courtesy History of Science Collections, University of Oklahoma Libraries; copyright the Board of Regents of the University of Oklahoma)

Copernicus undertook to improve upon predictive accuracy in astronomy, at first on a Ptolemaic basis, but gradually turning his attention to the alternative model offered in antiquity: the possibility of a moving earth. He sketched his system in the *Commentariolus*, an unpublished outline widely circulated in draft form among his students before 1514. Challenging established tradition, he proposed three kinds of terrestrial motion: (1) real diurnal rotation of the earth to account for the apparent diurnal rotation of the heavens; (2) annual revolution about the stationary sun to account for the solar year; and (3) motion in declination to account for the precession of the equinoxes.

Copernicus continued to elaborate his planetary theory during the next quarter of a century. In 1539 a Lutheran scholar from Wittenberg named Georg Joachim Rheticus (1514–1574) learned of the Catholic astronomer's theory and traveled to Frombork to study it in detail. Rheticus became Copernicus' first disciple, publishing in 1540 his own sketch of the system entitled *Narratio Prima*. He finally persuaded Copernicus to offer his theory to the world, and the latter authorized Rheticus to carry a copy of the manuscript to Nuremberg in 1541 where it was published by Johannes Petreius under the title *De revolutionibus orbium coelestium libri sex* ("Six books on the revolutions of the heavenly spheres"). As circumstances did not permit Rheticus to remain in Nuremberg to oversee publication, that duty was entrusted to the Lutheran theologian Andreas Osiander, who added an unauthorized preface stating that "[t]hese hypotheses need not be true nor even probable; if they provide a calculus consistent with the observations, that alone is sufficient..." (Gingerich, 2004:138–139). Osiander emphasized the hypothetical nature of astronomy used as a calculating device, apparently for the purpose of protecting the work from over-zealous censors. But whether or not Copernicus was aware of this preface when *De revolutionibus* was presented to him as he lay dying in 1543, he almost certainly would not have agreed with Osiander's disclaimer that heliocentrism should be treated only as a mathematical convenience, rather than as a genuine claim about the true physical nature of the cosmos.

THE CONSERVATIVE ORIGINS OF A REVOLUTION

Copernicus' *De revolutionibus* sits in the paradoxical position of being on the one hand essentially a conservative work in the classical astronomical tradition, and on the other hand a book that sparked a major revolution in scientific thought. With the exception of a broad exposition of the heliocentric system in the first of its six books, *De revolutionibus* is a highly mathematical treatise that made few initial converts. Although it was widely read in astronomical circles, fewer than a dozen committed Copernicans can be identified before 1600 (Westman, 1986:85). To preserve the assumption of uniform circular motion, Copernicus continued to employ Ptolemy's idea of epicycles and eccentrics, and has sometimes been referred to as the last great Ptolemaic astronomer.

Where Copernicus departed from the tradition of Ptolemy in a profoundly significant way, however, was in his pursuit of the insight that shifting the reference frame from the earth to the sun not only increased observational accuracy, but for the first time made logical sense out of the order of the planetary bodies. Rather than the sun, moon, and planets with their varying dimensions having been arbitrarily assigned to widely divergent periods and orbital angles, a heliocentric system generated an intrinsic order. The planets farthest from the sun had the longest orbital periods and the widest orbital angles, while those closest to the center revolved most tightly and rapidly around the sun. Likewise, the Copernican model also made coherent sense of retrograde motion. Instead of interpreting the looping paths of the planets against the sidereal backdrop as being actual celestial occurrences, Copernicus understood these motions to be mere optical illusions resulting from the annual revolution of our terrestrial observatory inside or outside the orbits of its fellow planets. Offering a remarkably prescient rebuttal to Ptolemy's objection that a moving earth would leave any loose objects drifting westward, Copernicus suggested two possible explanations. One was based on an Aristotelian mingling of qualities, the other on the idea of momentum: "The reason may be either that the nearby air, mingling with earthy or watery matter, conforms to the same nature as the earth, or that [this] air's motion, acquired from the earth by proximity, shares without resistance in its unceasing rotation" (Copernicus, 1992:I.8).

Initial reaction to Copernicus' revolutionary postulate was guarded, although astronomers appreciated the increased predictive accuracy of his system. More significantly, the fruitfulness of his effort may better be measured by the range and diversity of theories he stimulated. *De revolutionibus* gave free rein to an incremental rethinking of astronomy and physics that challenged the existing hierarchy of disciplines, and that within a century would blossom into a full-scale scientific revolution. Ptolemaic astronomy no longer offered a satisfactory architectonic vision of the cosmos, and Copernicus was not the only thinker prepared to suggest an alternative model. The Danish astronomer Tycho Brahe (1546–1601) proposed a "geo-heliocentric" model in which the five planets revolve around the sun, which in turn revolves with the moon around the earth. Brahe appreciated Copernicus's success in circumventing the most discordant aspects of the Ptolemaic system, but he personally could not overcome a revulsion of ascribing to the "sluggish earth" the quick motion shared by the "ethereal torches" (Dreyer, 2004:167–168). But Brahe did initiate a break with another important Aristotelian assumption—the immutable nature of the

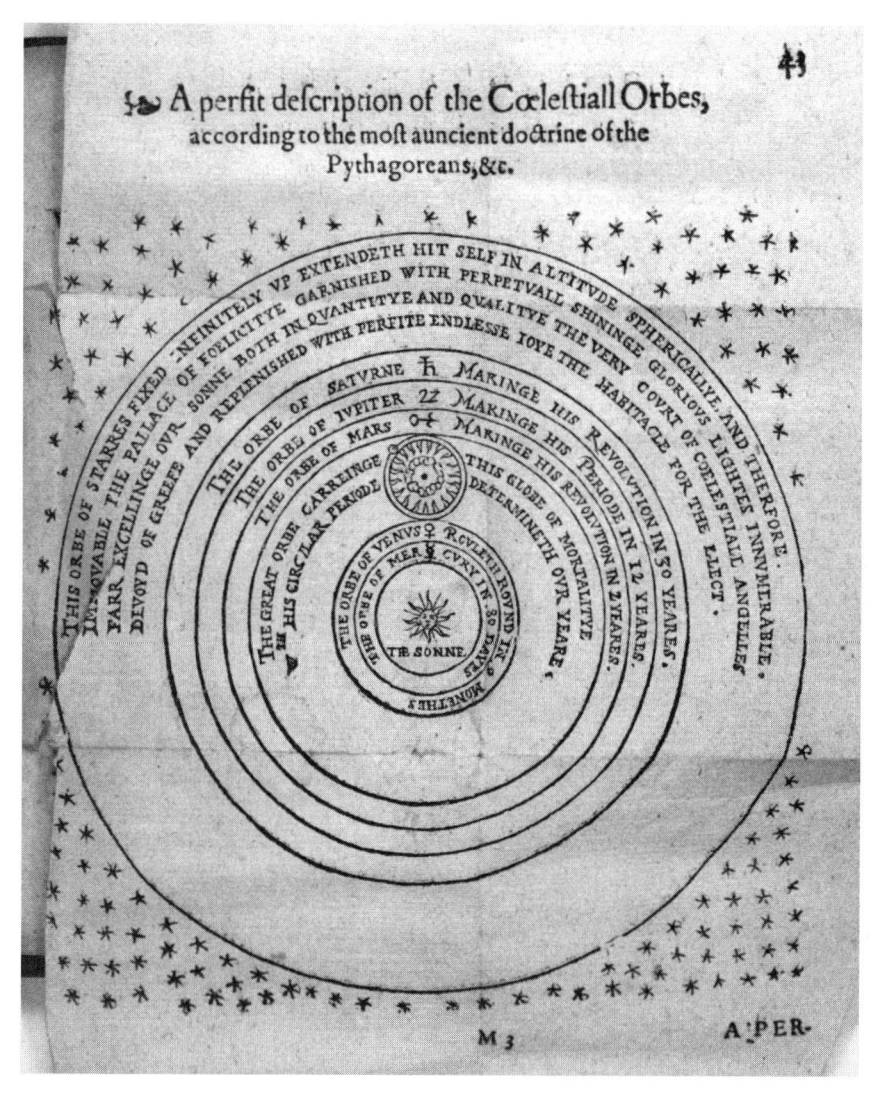

Thomas Digges (1546–1595). "A Perfit Description of the Caelestiall Orbes according to the most aunciente doctrine of the Pythagoreans, latelye revived by Copernicus and by Geometricall Demonstrations approved." First illustration of the Copernican model published in English, in a new edition his father Leonard Digges's almanac, *A Prognostication Everlasting* (1567). (Image courtesy History of Science Collections, University of Oklahoma Libraries; copyright the Board of Regents of the University of Oklahoma)

celestial spheres. First, against the Aristotelian dictum that the heavens are immutable, he claimed that the nova observed in 1572 was in fact a new star (*stella nova* is Latin for "new star") rather than a closer object, because it exhibited no parallax, that is, no progressive displacement against the background of "fixed stars." Second, because he observed that the comet of 1577 undeniably looped around the sun in an orbit closer than that of Venus, Brahe concluded that Aristotle's solid crystalline spheres could not exist.

Both the Catholic Copernicus and the Lutheran Brahe inhabited a pre-Newtonian age of relatively static astronomy, and their respective modifications of geocentrism remained committed to uniform circular motion as an attribute of perfection proper to the heavenly sphere. A more remarkable departure from classical astronomy was initiated by Brahe's student Johannes Kepler (1571–1630), whose close observation of Mars led him to postulate elliptical planetary orbits, with the sun occupying one focus of the ellipse. This "breaking of the circle" worked arguably an even greater psychological impact on the Western mind than did the shift to heliocentrism, for it destroyed the medieval ideal of celestial perfection (Nicholson, 1960:123–165). Furthermore, the introduction of a dynamic element with the planets revolving about a central sun was—even absent a theory of gravitation—a significant step on the path to an eventual Newtonian synthesis.

HUMANISM AND THE FOUNDATION OF THE LYNCEAN ACADEMY

The fortunes of the Copernican hypothesis were shaped not only by its incremental acceptance by natural philosophers, but also by factors such as the flexibility of intellectual culture and the circumstances of ecclesiastical politics. For a thousand years, the Catholic Church had been a unifying cultural force in Europe, helping establish the universities, supporting the arts through papal or episcopal patronage, and participating in natural philosophy through the work of its clerical scholars. But churches are not novelty-seeking institutions, and Catholicism was now facing serious internal and external challenges. It is important to bear in mind that however important a value it is in our twenty-first century perspective to preserve the autonomy of astronomical and other sciences, it would be quite anachronistic to expect to find this value prized in 1600.

To be sure, after the Condemnations of 1277 the Church had wisely refrained from committing itself to any particular cosmology. In his commentary on Aristotle's *De caelo* (1377) Bishop Nicole Oresme (1323–1382) had felt at liberty to consider a number of arguments for diurnal rotation, although ultimately he concluded in favor of the Aristotelian-Ptolemaic model. And in the relatively unconstrained atmosphere of scientific thought in the early sixteenth century Copernicus dedicated his book to Pope Paul III himself. However, although the immediate response to *De revolutionibus* on the part of the post-Tridentine church was to impose no particular restrictions on Catholic astronomers, not surprisingly this attitude would change as positions became more polarized in the following decades.

The scientific climate in the late sixteenth century was powerfully influenced by the Renaissance, that great awakening of Europe to renewed confidence in human capacity. Renaissance cosmologies were shaped by renewed interest in the Neo-Platonism of late antiquity (Grafton and Siraisi, 1999:16), and new forms of Platonism offered a variety of alternatives to scholastic Aristotelianism (Gaukroger, 2006:88–101). The rediscovery of perspective influenced more than art and architecture: music, history, the sciences, and other disciplines benefited from the reflective distance it offered. Humanism sharpened the critical faculties of historians such as Lorenzo Valla (1406–1457) who exposed the "Donation of Constantine"—a document in which Emperor Constantine had ostensibly donated the Western Roman Empire to the Catholic Church—as an eighth-century forgery. It was the new sense of perspective that enabled classical scholar Desiderius Erasmus (1466–1536) to advocate religious reform from within by satirizing the excesses of the late medieval church, as in his humorous colloquy "The Shipwreck" (Erasmus, 2007:315–326).

In like manner, humanist scholarship quickened the pulse of interest in science, both amateur and professional. In 1603 Federico Cesi, the visionary son of the powerful Duke of Aquasparta, founded with his friends in Rome the *Lyncaeorum Academia*, or the *Accademia dei Lincei* (Academy of the Lynxes). Taking the sharp-eyed lynx as their emblem, the *Lincei* were the first exclusively scientific body in the world, including among their colleagues the Dutch physician John Heck, the polymath Francesco Stelluti, and of course Galileo Galilei (Martini-Bettòlo, 1986:6–10). The group achieved international recognition with the publication in 1628 of *Rerum Medicarum Novae Hispaniae Thesaurus*, which was both a pharmacopeia of newly discovered Mexican drugs and an encyclopedic review of European scientific knowledge as produced by the experimental method. However, the academy was rather short-lived. Although scholars from across Europe (including Francis Bacon) were eager to be associated with it, the *Lincei* did not long survive the death in 1630 of Cesi, its founder and patron. This was unfortunate both theoretically for science and practically for the outcome of the Galileo affair. The last issue of the *Thesaurus* appeared in 1651, and Italian scientific collaboration no doubt suffered as a result of this loss. Apart from a desultory attempt to revive it in 1745 by a group of scientists in Rimini, on the Adriatic Coast, the *Accademia dei Lincei* fell from view until its brief resuscitation in 1800 and its eventual transformation into the Pontifical Academy. On the practical side, Federico Cesi had been a close friend and colleague of Galileo, and the latter's trial of 1632 might well have proceeded differently—or even been avoided altogether—had the moderating influence of Cesi's presence and the collective prestige of the Lincei been a strongly felt presence (Martini-Bettòlo, 1986:10).

The Renaissance papacy included among its numerous and sometimes extravagant expenditures the steady support of astronomy. Pope Gregory XIII (1502–1585) established what would eventually become the Vatican Observatory, erecting a tower in the Vatican buildings and causing it to be outfitted with the latest astronomical instruments. Among the first orders of business was to address a problem confronting both the Church and natural philosophy: the Julian calendar was by the sixteenth century four days out of phase with the lunar calendar. Pope Gregory established a committee to study the scientific data and implications involved in calendar reform, drawing upon the expertise of Christoph Clavius, a Jesuit mathematician from the Roman College and one of the most eminent astronomers of his day, who explained the need for the reform and expounded upon its nature (Lattis, 1994:20–22). A meridian line was installed in the observatory to assist astronomers, and the reform was carried out in 1582, resulting in the Gregorian calendar we follow today (Heilbron, 1999:36–46).

But although late scholastic and Renaissance intellectual culture may have fostered creativity in astronomical thought, the ecclesio-political climate of Counter-Reformation Europe was less favorable to the promotion of Copernicanism. After Trent—preoccupied as it was with defending doctrinal orthodoxy against Protestants—the Catholic Church was not about to embrace innovations that would undermine its dogmas, and Copernicus' heliocentric theory fell into this category. In both Protestant and Catholic contexts Aristotelianism persisted well into the seventeenth century, becoming a conservative touchstone as orthodoxies hardened (Schmitt, 1983:25–29). Moreover, astronomical speculation could carry unwelcome implications as in the case of Giordano Bruno, who postulated a plurality of worlds and the infinity of the universe (Yates, 1991:360ff). Although executed for theological heresy rather than for his scientific views, Bruno was a vehemently anti-Aristotelian admirer of Copernicus, and from the 1590s astronomical innovation became associated with heterodoxy. In the decades after the Council of Trent the Catholic Church entered what has been referred to as a period of "restrictive orthodoxies" in which Aristotelianism—which had lost the vitality it imparted to scholastic natural philosophy—began to be applied almost as a mechanical criterion of the truth (Shea, 1986:114–115). In such an atmosphere it is not surprising that the initially favorable reception of Galileo's telescopic discoveries should have been accompanied by suspicion of his

Copernicanism, and ultimately by the suspension of *De revolutionibus* in 1616 by the Congregation of the Index, "until corrected." This suspension was honored largely in the breach, with only about 8 percent of the 500 extant copies of the first edition censored by their owners in full compliance with the Index (Gingerich, 1981:45–61).

FOUNDATIONS OF THE MODERN WORLDVIEW: GALILEO

Into this context stepped Galileo Galilei (1564–1642), a proud Italian and faithful Catholic, mathematician and experimenter, a man often honored with the title "father of modern science." Galileo's historic encounter with church officials in Rome is the incident most often adduced as an archetype of the "conflict model," shorthand for an assumed state of persistent warfare between courageous scientific pioneers and obscurantist ecclesiastical authorities. A cascade of recent historiography confirms, however, that a simple conflict model is essentially useless for penetrating beneath the surface to the profound intellectual, scientific, theological, cultural, professional, and personal issues intertwined in this famous episode. It is historically quite naïve to criticize Galileo's opponents for failing to accept his theory:

The new science, which today pervades our entire life, was just emerging, and very few were able to realize what was happening at the time. Most people were not ready to abandon cherished traditional ideas for daring hypotheses that had yet to be proved. (Artigas, 2003:ix)

Galileo in fact in his lifetime did not find the proof he needed to demonstrate conclusively that the earth revolved around the sun. Furthermore, incomprehensible though it may be to us now that the Catholic Church should have possessed any authority to suppress discussion of a scientific theory, we must bear in mind the role the Church had played for a thousand years after the fall of the Roman Empire in keeping intellectual culture alive in Christendom. As founder and supporter of institutions of higher learning, and as interpreter of the canon of scripture that Christians regarded as vital to their salvation, churchmen took seriously their responsibility to protect the faithful by guarding the deposit of faith. In 1600 the medieval vision of the unity of truth was alive and well, and truth in astronomy was quite relevant to truth in theology.

The "Galileo Affair": Résumé

Born in 1564 in Pisa, Galileo studied at the University of Pisa where he developed an early interest in mathematics and its application to the

Galileo Galilei (1564–1642). Frontispiece from *Sidereus Nuncius*, 1610. (Image courtesy History of Science Collections, University of Oklahoma Libraries; copyright the Board of Regents of the University of Oklahoma)

physical world, and where he followed Archimedes rather than Aristotle in certain problems in physics. He began teaching mathematical subjects at Pisa in 1589, obtained the chair of mathematics at the University of Padua in 1592, and as early as 1595 revealed his incipient Copernicanism by developing his own erroneous theory of the tides. Over the next fifteen years he worked on the problem of falling bodies and directed the first astronomically useful telescope toward the heavens, publishing his discoveries about the Moon and Jupiter in his book *Sidereus Nuncius* in 1610. With the publication of his telescopic observations physical evidence became available that seemed to confirm the mathematical theory of heliocentrism, although genuine empirical proof of Earth's annual orbital motion would only arrive with Bessel's establishment of stellar parallax in the 1830s, and the confirmation of diurnal axial rotation awaited Foucault's pendulum in 1843. Nevertheless, Galileo's charting of the revolutions of the moons of Jupiter and the phases of Venus suggested, by analogy, the plausibility of the heliocentric cosmological model. Likewise, his observation of lunar craters implied the similarity of the moon to the earth, and his discovery of sunspots furthered the argument that mutability is not confined only to the terrestrial realm.

Galileo was rewarded with lifetime appointment to the post of "Chief Mathematician of the University of Pisa and Philosopher and Mathematician to the Grand Duke of Tuscany" and moved to Florence. He began to make visits to Rome, where he was celebrated by the Jesuits as a discoverer. The first hint of trouble came in 1613 with the publication of his *Letters on Sunspots*, and the problems accelerated two years later with the circulation of his "*Letter to the Grand Duchess Christina.*" In this publicly circulated "letter" Galileo laid out the principles of biblical interpretation that undergirded his sincere belief that heliocentrism was not in conflict with any scriptural passages that touched upon cosmology.

I think that in the discussion of natural problems we ought to begin not with the Scriptures, but with experiments and demonstrations. Nor does God less admirably discover Himself to us in nature than in Scripture, and having found the truth in nature we may use it as an aid to the true exposition of the Scriptures. The Scriptures were intended to teach men those things which cannot be learned otherwise than by the mouth of the Holy Spirit; but we are meant to use our senses and reason in discovering for ourselves things within their scope and capacity, and hence certain sciences are neglected in the Holy Writ. (Galileo, 1957:182)

Galileo argued that in those few biblical passages referring to the mobility of the sun (Joshua 10:12–13, Psalms 19:4–6, Ecclesiastes 1:5) or to the stability of the earth (1 Chronicles 16:30, Psalms 93:1, 2 Samuel 22:16, Psalms

OBSERVAT. SIDEREAE

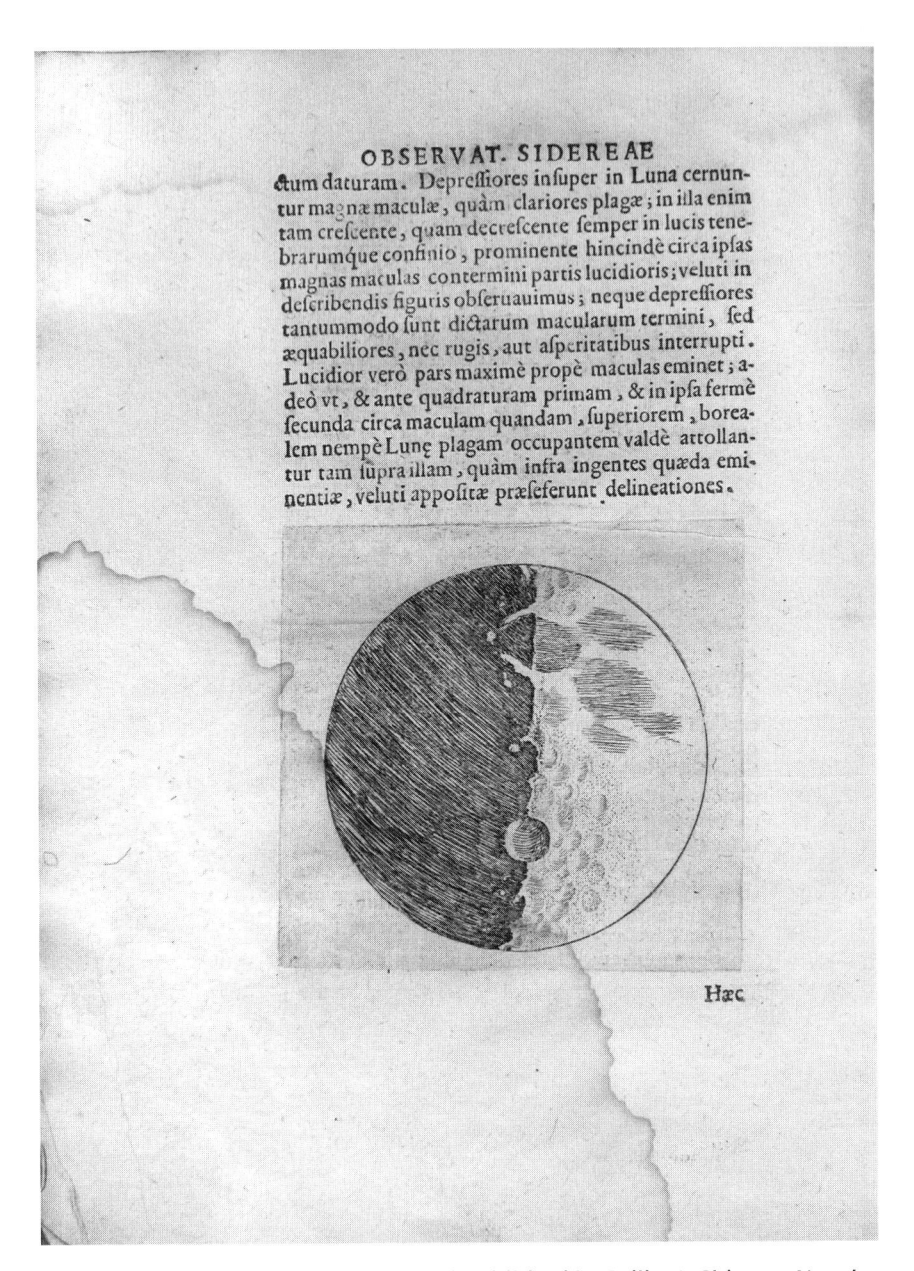

&tum daturam. Depressiores insuper in Luna cernuntur magnæ maculæ, quàm clariores plagæ; in illa enim tam crescente, quam decrescente semper in lucis tenebrarumque confinio, prominente hincindè circa ipsas magnas maculas contermini partis lucidioris; veluti in describendis figuris obseruauimus; neque depressiores tantummodo sunt dictarum macularum termini, sed æquabiliores, nec rugis, aut asperitatibus interrupti. Lucidior verò pars maximè propè maculas eminet; adeò vt, & ante quadraturam primam, & in ipsa fermè secunda circa maculam quandam, superiorem, borealem nempè Lunę plagam occupantem valdè attollantur tam supra illam, quàm infra ingentes quæda eminentiæ, veluti appositæ præseferunt delineationes.

Hæc

Illustration of lunar topographic detail published in Galileo's *Sidereus Nuncius* (1610). (Image courtesy History of Science Collections, University of Oklahoma Libraries; copyright the Board of Regents of the University of Oklahoma)

104:5, Job 38:4–6), the biblical authors were writing figuratively. They were merely accommodating their language to the understanding of simple and illiterate people (Blackwell, 1991:53–86). The principle of accommodation would play an increasing role not only in discussions about astronomy, but also in eighteenth-century debates about the age of the earth.

Cardinal Robert Bellarmine cleared Galileo's name at this stage, but controversy continued with Galileo's anti-Aristotelian interpretation of the three comets observed in 1618; Aristotle had held them to be atmospheric phenomena (*Meteorology*, I.7; Genuth, 1997:17–19). Shortly thereafter, the Copernican theory was condemned, although the Congregation of the Index permitted scholars to read *De Revolutionibus*, providing it was understood hypothetically and that certain minor corrections were made.

Despite this, Galileo was still on good terms with Rome in 1624, and was warmly welcomed by Pope Urban VIII. However, with the completion of his *Dialogue Concerning the Two Chief World Systems* five years later, it appears that various forces opposed to Galileo's gradual challenges to the Aristotelian-Ptolemaic system were provoked beyond the limits of tolerance by certain factions within the Church. In 1632 Galileo was ordered to Rome to stand trial, where his theories and the teaching of them were condemned, and where in June 1633 he capitulated. The substance of the abjuration is as follows: (1) The proposition that the sun is in the center of the world and immovable from its place is absurd, philosophically false, and formally heretical; because it is expressly contrary to Holy Scriptures; (2) The proposition that the earth is not the center of the world, nor immovable, but that it moves, and also with a diurnal action, is also absurd, philosophically false, and, theologically considered, at least erroneous in faith.

For the remaining ten years of his life, Galileo was confined to Florence under house arrest, publishing his most famous work, the *Discourse on Two New Sciences* in 1638 in Calvinist Holland. He died in 1642, attending Mass almost until the end. Galileo's *Dialogue* was removed from the Index nearly two centuries later, at last permitting Catholics to teach Copernicanism with no qualification (Shea, 1986:132).

The "Galileo Affair": Issues of Contention

The controversy surrounding Galileo's publicly maintaining a heliocentric interpretation of his telescopic discoveries has come down to us as the archetype of the age-old conflict between science and religion (White, 1897:177–199). But a careful reading of the historical evidence suggests (1) that the points of conflict were multiple rather than reflecting a simplistic dichotomy of "scientists vs. churchmen," (2) that where cosmology was at stake it was an issue of conflict between conflicting worldviews, not between science and religion, and (3) that a number of points of conflict had to do with matters other than cosmological questions. We shall treat this complex of problems under four headings: intellectual, cultural, ecclesiastical, and personal issues.

Intellectual Issues As noted above, however obvious the physical truth of the Copernican cosmological model may seem now to us, it was by no means obvious to Galileo's contemporaries; it contradicted common sense. Mathematically it may have been more elegant to eliminate most of Ptolemy's epicycles, but heliocentrism was not self-evidently true according to the principles of seventeenth-century natural philosophy (Popkin, 2002:90). Although the moons of Jupiter and the phases of Venus suggested by analogy the truth of a heliocentric cosmological model, an empirical proof of Earth's annual revolution about the Sun would not be available for more than two centuries, until the discovery of stellar parallax.

Intellectual conflicts also arose over epistemological issues, such as Galileo's elevation of sense experience—in the form of experiment and observation—into a far more exalted role than it had enjoyed in Scholastic Aristotelianism. Moreover, Galileo's insistence on the physical truth of Copernicanism—despite his lack of conclusive proof—suggests his unorthodox willingness to accept as true in natural philosophy a proposition that is not yet known with certainty.

In addition, Galileo elevated mathematics to the status of the divine language of nature, and upset the established order of the scholastic curriculum by daring to use a lower science, mathematics, to correct astronomy, a science of higher dignity. It has been argued that that the Renaissance fascination with the significance of number in Platonic thought was one of the solvents of Aristotelianism in the sixteenth century, supporting a scientific counterculture of which Galileo became a part with his desire for the "mathematization of the cosmos" (Hankins, 1999:88). In the late Renaissance, non-Aristotelians idealized mathematics as reflecting the spiritual nature of the universe (Bouwsma, 2000:189). The Lutheran astronomer Kepler held the rather esoteric belief that God had created the universe according to a precise mathematical plan, in which the diameters of the orbital spheres of the planets corresponded to the five regular solids. Accustomed as we are in the twenty-first century to the intrinsically mathematical nature of physics and astronomy, we must bear in mind that in Galileo's time this was still a novel and upsetting idea.

Intellectual conflicts arose on the theological side over a variety of issues ranging from Galileo's promotion of atomism (sec. 2.7) that threatened the Scholastic doctrine of transubstantiation in Eucharistic theology to the question of who possessed authority to interpret the Bible regarding what it says about nature (Langford, 1992:50–78). With respect to the exegesis of scriptural passages implying the mobility of the sun or the stability of the earth, Galileo's views were not incompatible with those of some early church fathers (Dorlodot, 1922:169).

Cultural Issues Second, the Galileo affair must be considered in the wider historical context of sixteenth- and seventeenth-century Italy. William Shea has pointed out that it is a mistake to interpret the Counter-Reformation as a cause of Rome's retrenchment to orthodoxy. Rather, the Counter-Reformation was merely one symptom of what he refers to as a "crisis of confidence" in the Italian mind. The loss of half of Christendom to Protestantism, the collapse of the Florentine Republic, and the subsequent domination of the Italian peninsula by Spanish troops caused a psychic trauma for the Church in Rome so great that it reacted by generally putting the brakes on novelty, including scientific innovation. Aristotelian thought—which only a few centuries before had been so liberating—became for the Church "a mechanical criterion of truth" (Shea, 1986: 114–16).

Galileo's misfortune was in part that his brilliant scientific career flowered both temporally and geographically in the very midst of Rome's powerful reaffirmation of the Catholic tradition following the Protestant Reformation. While the Counter-Reformation was not in itself antiscience, it was certainly not about to embrace innovations that would undermine its dogmas (Ashworth, 1986:148–149). One consequence of this was the papal mandate to carry out the policies of the Council of Trent, including an exegesis that emphasized wherever possible the literal interpretation of the Bible. Furthermore, Galileo as a layman would not have been entitled to expound upon the scriptures, and when he did so in his *Letter to the Grand Duchess Christina* he was treading on very dangerous ground indeed.

To this should be added the conflict within academic culture over professional "turf." Peculiar though it may seem to us in an era of autonomous disciplines sharing a roughly equal academic playing field, in Galileo's time theology had reigned for centuries as "Queen of the sciences." Astronomy was part of the quadrivium, a distinctly inferior discipline, and the fact that astronomers were aspiring to greater autonomy and importance rankled the establishment. Moreover, Galileo's involvement with and support by Federico Cesi's short-lived *Accademia dei Lincei* may have been viewed as vaguely threatening, for the *Lincei* was the first scientific society established with no direct ties either to church or to conventional academe. Yet another challenge to the established order was Galileo's reaching out to the interested lay public by scandalously publishing in Italian, not exclusively in Latin, the language of scholars.

Issues of Political and Ecclesiastical Conflict Galileo unwittingly or naively became embroiled in ecclesiastical intrigues between Florence and Rome, and between Dominicans and Jesuits and Carmelites (Feldhay, 1995:171–198). It would not serve our purposes to rehearse the intricacies of the proceedings against Galileo, as these have received numerous excellent treatments (Artigas and Shea, 2003; Lindberg, 2003a; Fantoli, 2003). However, it is essential to underline the significant internecine rivalries within and between various Jesuit and Dominican factions in the Roman Church that had nothing to do with Galileo, and about which he may have been largely ignorant).

It has been argued that the Galileo affair was in some respects an event local to the Italian peninsula. Of course, in light of the consequences of the trial of 1633 for Galileo personally, one hesitates to underestimate the power of the Holy Office. However, the Inquisition's effectiveness was not uniform throughout Europe, and the historian Owen Gingerich has shown that the censorship of Copernicus' *De Revolutionibus* was little honored outside Italy (Gingerich, 2002:xxiv). Still, as we shall see below, the episode did have a chilling effect on science within Italy (Ashworth, 1986:148–153). And it was not only the Inquisition that could impose internal conflict: although Galileo's Jesuit colleagues who were competent mathematicians suspected that heliocentrism was correct, they were compelled by a directive from their own order to teach an account of the universe as a set of finite concentric spheres rotating above a stationary earth (Rowland, 2000:2; Fantoli, 2003:106–109).

Personal Issues Finally, there were significant personal factors that contributed to the collision between Galileo and the Church at Rome. While it is indisputable that Galileo outclassed other astronomers in the brilliance of his theorizing, regrettably he lacked the humility to bear his talent with grace. His treatment of the discoveries of other astronomers as inferior to his own deprived him of the benefit of their insights. For example, he rejected Kepler's discovery (communicated to him by Cesi) that assuming orbits to be elliptical allows for greater accuracy in calculating planetary motion. Galileo was convinced that only perfectly circular orbits could permit natural and unending motion (Artigas, 2003:32, 54). Galileo's assertive and arrogant style in both debate and writing insulted and alienated not a few people who had been his friends and supporters, including Maffeo Barberini, later Pope Urban VIII. In an astonishing blunder, Galileo in his *Dialogue concerning the Two Chief Systems of the World* (1632) put into the mouth of Simplicio ("the simpleton") the words of his former friend: "It would be excessively bold if someone should want to limit and compel divine power and wisdom to a particular fancy of his."

Galileo's own struggle between religious faith and his confidence in science mirrors the intellectual ferment of the age. Whatever their confessional commitments, most (though not all) scientists in 1630 still demonstrated some degree of genuine loyalty to either Protestant or Catholic Christianity. Galileo's acceptance of adversity through his trials shows his commitment to science and his deep religious belief. In contrast with some of his opponents in Rome for whom ecclesiastical positions were principally the rungs of a career ladder (Spini, 1966:66), Galileo comes across as a man of genuine faith. As he remarked in the beginning of the *Dialogue concerning the Two Chief Systems of the World*, he was motivated in part by his desire to save his Church from error. Whether or not we accept the sincerity of this literary declaration, it cannot be denied that Galileo appears to have remained a loyal Catholic until his death.

To summarize the Galileo episode, we have seen that underlying this spectacular conflict are a host of factors that render it impossible for us to interpret it as exemplifying the conflict model between science and religion. Moreover, it is also important to note that at least two of Barbour's other models are operative here (Barbour, 1997:77–105). In the *Letter to the Grand Duchess Christina*, Galileo's quotation of Cardinal Baronius's adage that "scripture teaches how the heavens go, not how we go to heaven," might appear to support an independence model. But in the case of Galileo this would be inaccurate, because the general tenor of his life and thought suggests that dialogue between the two is quite important, and perhaps even that he strove for integration of his science and his religion into a coherent whole.

CATHOLICISM AND SCIENCE IN THE AFTERMATH OF GALILEO

The Galileo affair often assumes such prominence in many historical accounts that it obscures the fact that Catholics had practiced science for hundreds of years before and would continue to do so after this largely political affair. Catholic thinkers on both sides of the controversy over the heliocentric hypothesis were committed to the question of truth in both cosmology and theology. Scientific work among Catholics did not stop, but it did suffer some diminishment both geographically and disciplinarily. Just what were the consequences for Catholics working in science, and for its relationship to theology or religious practice?

In a provocative essay, William Ashworth considered why major Catholic contributions to science were rare in the later seventeenth century. Reviewing the careers of five major Catholic figures, he concluded that no discernible pattern of "Catholics science" emerges from a survey of the life and work of Marin Mersenne, René Descartes, Pierre Gassendi,

Blaise Pascal, and Nicholas Steno. Between them are represented empiricism, rationalism, skepticism, mysticism, mechanical philosophy, natural theology, atomism, and mathematism (Ashworth, 1986:147). However, although no personal qualities distinguish this group of scientists from their Protestant colleagues, taken as a whole Protestant countries produced more science than Catholic countries after 1630. Ashworth argued that the outcome of Galileo's second trial was quite sobering for Catholic scientists, particularly in Italy, and that what made the difference was the vast bureaucracy of the Tridentine Catholic Church and the new tools at its disposal, the Inquisition and the Index. No other denomination in Europe was as well equipped for the enforcement of orthodoxy. The simple fact was that "Newton, Boyle, Leibniz, and Descartes spun off hypotheses in profusion, and science flourished in their wake. In Italy after 1650, a hypothesis could hardly be found, and by 1700 science was struggling to survive" (Ashworth, 1986:148–153).

The Society of Jesus grew to be an enormously important force in the sphere of Catholic education following the Council of Trent. As a teaching order its members were in a prime position to make significant contributions to many areas of science and mathematics, such as geometry (MacDonnell, 1989:1–5). The flamboyant Athanasius Kircher (1602–1680) was brought to Rome in 1635 (only two years after Galileo's condemnation) to fill the chair of mathematics in the Jesuits' flagship institution, the Collegio Romano. Catholic orthodoxy needed a figure whose glamour approached that of Galileo (Rowland, 2000:2), and Kircher was a brilliant choice. He installed his astonishing collection of human, geological, and natural historical artifacts in what he called his *Musaeum* in the Collegio Romano, a veritable showcase of the Catholic investment in science (Rowland, 2000:3–6).

Although no one doubts the breadth of the Jesuit pedagogical vision, historians are divided as to the impact of their contribution to the early modern scientific enterprise, and as to why there were no Jesuits of the stature of the five scientists mentioned in the previous paragraph. Ashworth has claimed that Jesuits allowed their missionary concerns to shape their epistemology and view of nature, isolating their scientific work from the mainstream of the Scientific Revolution. In particular, he singled out their "emblematic view of the world," in which nature is viewed primarily as a hieroglyph, a vast collection of signs and metaphors (such as "Christomorphic" stones) employed for the purposes of teaching in a missionary context, where the question of truth about natural things is irrelevant. Second, he suggested that the Jesuits embraced a form of Probabilism on matters of fact, not just of morality. Every respectable opinion carries the same weight as any other, a position that prevented them from developing a consistent natural philosophy. Giovanni Battista Riccioli (1598–1671), for example, the first person to measure the rate of acceleration of a freely falling body (even before Galileo), offered in one of his works an erudite discussion of new stars. He enumerated fourteen possible causes of these phenomena, but in the end did not commit himself to a single one, not even to his own, as being more important than the others in that it tells us something meaningful about the cosmos. The third deficiency of Jesuit science in Ashworth's view is that in the wake of the Inquisition's edict against Galileo in 1616, the Jesuits adopted the principle of "fictionalism": they were permitted to discuss heliocentrism (or another non-Aristotelian theory) with enthusiasm and erudition, but could not hold it to be more than a fiction. This was a recipe for obsolescence in a scientific community in which Boyle was formulating nonfictitious laws about gases and Newton was publishing his laws of motion that he took to be genuine (Ashworth, 1986:156–160).

The range and quality of Jesuit participation in the sciences were vast. Mordechai Feingold and others have offered a more positive and carefully nuanced appraisal of the Jesuit contribution to science in *Jesuit Science and the Republic of Letters* (2003). Recognizing the frequently dismissive or patronizing tone of evaluations of Jesuit science, Feingold contends that members of the society who zealously engaged in scientific studies were often blocked by the impediments placed in their way by their superiors or by fellow Jesuits. Such obstructions ranged from the order's ambiguous attitude toward publication (did Jesuit books glorify God or the author?) to the confiscation and even vindictive destruction of scientific apparatus. Most important, the mission of the Society of Jesus was to save souls, not to work at the cutting edge of science. Still, even educating future scientists and scholars was a contribution as fundamentally important as formulating scientific laws:

We may recall that the Jesuits produced Torricelli, Descartes, Mersenne, Fontenelle, Laplace, Volta, Diderot, Helvétius, Condorcet, Turgot, Voltaire, Vico, and Muratori, to name but a few non-Jesuits. (Feingold, 2003:38)

Nor were obstacles always the fault of the society or of individual Jesuits. The astronomical activity of members in the Spanish province was hampered by Spain's ideological isolation (imposed to preserve orthodoxy) that "acted increasingly as a barrier to new philosophical and scientific ideas" (Navarro, 2003:331). Nevertheless, it cannot be gainsaid that the commitment of the Society of Jesus to a particular interpretation of the "unity of truth" inevitably constrained its members' involvement in the early modern scientific enterprise:

In the modified Thomism the Order adopted, the term "science" was applied to a continuous doctrinal fabric, from the principles of metaphysics to the explanation of particular natural phenomena. In turn, the metaphysical principles were set down in strict correlation with the scholastic interpretation of Christian dogma. (Baldini, 2003:69)

Such an ideological context made it difficult for the most adroit of Jesuit scientific thinkers both to remain loyal to the spirit of their order and to engage at the highest level with the scientific community of discourse.

CATHOLICISM AND THE PHYSICAL SCIENCES IN THE SEVENTEENTH CENTURY

The perspectival shifts from geocentrism to heliocentrism and from circular to elliptical orbits were only part of the early modern story. Equally portentous for the relationship between Catholicism and the sciences was the development of a new corpuscularian physics, the roots of which are both physical and theological. In the fourteenth century, physics was still fundamentally qualitative, and was based upon the scholastic Aristotelian doctrine of the four elements: earth, water, air, and fire, and of a fifth essence or "quintessence" of which objects in the celestial realm were composed. However, a developing fourteenth-century fascination with measuring everything possible (Murdoch, 1975:287) coincided with a growing dissatisfaction with the ability of Aristotelianism to answer such questions as the nature of a hypothetical vacuum. Additionally, theological objections were raised against Renaissance naturalism, which drew upon the Aristotelian idea that nature is permeated with active principles, endowing it with autonomy. If such active principles existed, the line between natural and supernatural would be dangerously blurred, and the world would be alive with occult properties (Gaukroger, 2006:255). These fourteenth-century discussions in both theology and natural philosophy would lead to a revolution even more dramatic than that undergone by astronomy: the momentous shift from the Aristotelian worldview to the mechanical philosophy.

Philosophical mechanism offered a bridge between matter theory and the science of mechanics, which held motion to be the basic explanatory tool (Gaukroger, 2006:200). Mechanism is inherently mathematical. Niccoló Tartaglia (1500–1557) employed mathematics to calculate the trajectories of cannonballs in Venice, and the Jesuit Giovanni Riccioli used it to determine the rate of acceleration of falling bodies. From Kepler and Galileo to Descartes and Newton, scientists came to realize that understanding mathematics is fundamental to understanding nature. If the mechanical philosophy is about measuring matter in motion, it became clear that the Aristotelian conception of nature as composed of the four elements was untenable. But what was the alternative? Natural philosophers had to look with new eyes to see the alternative ready to hand. The theory of an atomic constitution of nature (*atomos* is Greek for "that which cannot be cut") was not new to the seventeenth century, having first been propounded by Leucippus and Democritus in the fifth century BCE and then systematically elaborated by Epicurus (341–270 BCE) and Lucretius (96–55 BCE). As an explanation of the fundamental structure of the world, the atomic hypothesis had been submerged in antiquity by the dominant Aristotelian physics of four elements. Moreover, a world created and governed by the chance collisions of an infinite number of atoms proved to be incompatible with the Christian doctrines of providence and free will, whereas Aristotle's system, when purged of the unacceptable doctrine of the eternity of the world, carried with it no threatening overtones of atheism.

Resuscitated at the same time that Aristotelian physics was losing favor, the reception accorded the atomic theory was guarded. "Epicureanism" in the sixteenth century was generally equated with the dreaded specter of atheism, and many theologians therefore blamed the rise of atheism explicitly on that "accursed philosophy." However Michael Buckley has argued convincingly that modern atheism owed its rise not to a particular worldview but rather to the seventeenth-century appeal to science for validation rather than to intrinsic religious experience (Buckley, 1987:341–362). As the revolution in scientific perspective progressed, it became clear that atomism was an immensely fruitful hypothesis, not inimical to religious belief, and able to account for much that was simply inexplicable on Aristotelian assumptions, such as the existence of a vacuum (Kargon, 1966:3–4; Trevor-Roper 1987:1–39). But cherished intellectual commitments are not easily relinquished, and fifty years after Evangelista Torricelli had demonstrated a vacuum, a scholastically trained English Catholic priest, John Sergeant, vehemently denied that it was possible:

In vain do these ingenious men endeavor to torture nature by suckers in pumps, and such inventions, to find out a vacuum. The essences of everything are established by the first being, and sooner may nature be torn into atoms, than they can cease to be what they are. (Sergeant, 1700:326)

This was not simply a gratuitous attack on Boyle's empirical investigations into the idea of a vacuum, but rather was intrinsically related to Sergeant's argument for the existence of God. In order to account for the transmission of motion from God to the smallest particles of the universe, he had to maintain that no vacuum could exist. Sergeant's natural theology and his conception of providence thus forced him to prop up a theory of physics increasingly disregarded by serious scientists.

In contrast to Sergeant, the French Franciscan Minim friar, Marin Mersenne (1588–1648), represented the future, arguing that the way to combat naturalism and mortalism (the denial of immortality) was to promote metaphysical mechanism, a system in which matter is completely inert (Gaukroger, 2006:256). But the Catholic thinker who most prominently campaigned for atomism was Pierre Gassendi (1592–1655), a Catholic priest opposed to Aristotelian science and to alchemy and similar natural philosophies. In Gassendi's view the system that was most compatible with the doctrines of creation and divine providence was a physics premised upon a finite number of inert atoms (Lucretius had held them to be infinite) under the providential governance of God (Gaukroger, 2006:262–276).

Gassendi's contemporary, René Descartes (1596–1650) likewise promoted a corpuscularian philosophy, but he drew out more daring implications and provoked greater theological displeasure. Descartes developed a cosmological "vortex theory" in which solar-planetary systems might be formed, torn asunder, and reformed by the collisions of particles swirling through the fluid vortices that fill the universe. The earth might have been formed from the cooled-off core of a wandering star that became trapped in the vortex of our own solar system. Descartes declared that he accepted the miraculous creation of the world recounted in Genesis 1, but that "in our imagination" the result would have been the same, and would account better for phenomena such as earthquakes and volcanic activity (Gaukroger, 2006:311–322).

Far more serious from the scholastic point of view were the theological objections that Descartes' matter theory provoked. The theology of "transubstantiation" (defined in 1215 at the Fourth Lateran Council) accounted for the "real presence" of Christ in the bread and wine of the Eucharist in terms of Aristotelian metaphysics: the substance of the bread and wine was believed to be changed into that of Christ's body and blood, while the accidents (color, taste, and texture) remained the same. Cartesianism unwittingly challenged this:

He [Descartes] failed to anticipate that by equating matter with extension and denying the reality of secondary qualities, he had completely undermined the doctrine of the Eucharist. If accidents have no real existence, how could the accidents remain while the substance was miraculously transformed into the body of Christ? (Ashworth, 1986:140)

The new atomic theory of matter, therefore, and the mechanical philosophy of matter in motion were not free of theological objections, any more than had been Aristotelianism. However, just as Aristotelian philosophy had been absorbed and to some extent controlled by medieval Catholicism, so atomism and the mechanical philosophy would be adapted to early modern theologies, albeit gradually and with qualifications (Deason, 1986:186–188).

About Catholicism and chemistry we shall say little, because few Catholics dared practice it after Trent, as it bore too close a relation to darker arts. The Church had long been afraid of magic (such books appeared in the first edition of *Index*), including divination, which was regarded as conferring demonic power upon the individual "magus," and astrology that threatened the doctrine of free will (Ashworth, 1986:149). Alchemy was not originally prohibited, perhaps because the great Renaissance alchemist Paracelsus was a Catholic (although a heterodox one). In 1630, however, the theological faculty at the University of Louvain convicted the alchemist Jan van Helmont of using magic and diabolical art to pervert nature, and other alchemists were persecuted as well, with the result that apart from Helmont there was no notable Catholic contribution to chemistry before 1700. Although Robert Boyle was engaged in serious experimentation with gases, and distinguished between chemistry and alchemy in *The Sceptical Chymist or Chymico-Physical Doubts & Paradoxes* (1661), the real development of chemistry began in the late eighteenth century. Prominent among experimental chemists was Catholic layman Antoine Lavoisier, although religious beliefs do not appear to have had a bearing on his science.

THE POPES AND SCIENCE IN THE NINETEENTH CENTURY

The Catholic Church's enduring interest in science is illustrated by an effort to support collaboration across disciplines. Early in the 1800s, under Pope Pius VII (1800–1823), a Roman priest named Feliciano Scarpellini attempted to resuscitate the Academy of the *Lincei*. As industrial technology gathered steam in European society, scientific research into electricity was fostering new systems of communication, and the science of chemistry was rapidly developing in sophistication. Scarpellini resolved to gather some of the most distinguished scientists working in the Papal States into a close and collaborative association. These included Fr. Domenico Chelini (mathematics), Carlo Bonaparte (natural history), Alessandro Flajani (anatomy), Domenico Morichini (chemistry), Pietro Peretti (pharmaceutical chemistry), and the physicists Gioacchino Pesutti and Paolo Volpicelli (Martini-Bettòlo, 1986:10).

The Papacy was in the course of the nineteenth century besieged with political problems in the nascent Italian republic. Nevertheless, the Church maintained its support of the sciences, in which Pope Pius IX (1792–1847) had an enduring interest: before he became pope, Giovanni Mastai-Ferreti had published on the science of optics as a student. Immediately after his election in 1847, he restored the languishing New Lincei officially as the Academy of Sciences of the Pontifical States. Its scope was the promotion of science and advising of the government about new technologies to be adopted, and even about the approval of industrial patents." (Martini-Bettòlo 1986:11; Sorondo at www.disf.org). An early strategic decision was to introduce the metric system into the papal states.

The Academy of Sciences was constituted to have thirty resident and forty corresponding members, whose task was to "promote the study, the progress, and the development of the sciences, with the exception of theology, moral sciences, medical sciences, and political sciences" (Martini-Bettòlo, 1986:11). A bold innovation was to elect the Countess Fiorini, professor of botany in Rome, as its first female member. Outstanding work was done by astronomers among the members of the academy, such as Fr. Angelo Secchi, professor of astronomy in the Jesuit College in Rome, and director of its observatory. Secchi catalogued more than 10,000 double stars and pioneered the study of stellar spectroscopy" (Walsh 1909:209–213). He measured the height of the lunar mountains and was the first to observe the canals on Mars. He also studied and photographed the solar corona in eclipse, wrote about sunspots, and published *Le Soleil* (1870) that served as a basic nineteenth-century treatise on solar phenomena. Secchi thought at more comprehensive levels too, however, seeking to integrate existing fields of scientific knowledge, and attempting in his *Sulla unitá delle forze fisiche* (1864) to trace all processes of nature back to kinetic energy (Pohle, 1912a). This was no reductionist enterprise, moreover, as Secchi's Christian theism proscribed his extension of "kinematic atomistics" into the realm of the intellect and the soul. Engaging in a sort of astronomical natural theology, Secchi found that his science only strengthened his belief in the creator and in providence.

Since the time of Pope Gregory XIII, under whose auspices the calendar had been reformed—and notwithstanding the Galileo episode—the popes had been interested in astronomical research. Early observatories of papal foundation were the Observatory of the Roman College (1774–1878), the Observatory of the Capitol (1827–1870), and the *Specula Vaticana* (1789–1821) located in the Tower of the Winds within the Vatican grounds. In conscious furtherance of this tradition of patronage, Pope Leo XIII (successor to Pius IX) reestablished the Vatican Observatory in 1891, on a hillside behind St. Peter's Basilica. (Vatican Observatory Web site). Due to increasing light pollution from Rome, the observatory was moved to Castel Gandolfo, the papal summer residence, and in the 1980s to Kitt Peak in Arizona. The first three directors represented three different religious orders: Francesco Denza was a Barnabite friar who initiated the observatory's work in stellar photography, Angelo Rodriguez a Dominican specialist in meteorology, and Johann Hagen an Austrian Jesuit. Hagen had migrated to America where he began what became a valuable *Atlas of Variable Stars* (1899) and taught in Wisconsin and at Georgetown University before being appointed director (Walsh, 1909:217–220; Faherty, 1941:25). More will be said in Chapter 4 about Pope Pius XI's renewal and reconstitution of the Academy in 1936, as the Pontifical Academy of Sciences.

CATHOLICISM AND PHYSICAL SCIENCE: THREE MODELS OF INTERACTION

In the medieval and early modern periods, Catholic astronomers and other physical scientists were able for the most part to conduct their work without apparent contradiction between their science and their religious belief. The eighteenth-century Enlightenment altered this dynamic significantly. Enlightenment philosophers advocated reason as the basis of authority, not tradition or private revelation, and they appealed for toleration in the wake of the bitter and devastating wars of religion. Some advocated deism, in which God became a semi-impersonal creator who had withdrawn from further interference in terrestrial affairs. About Catholicism and the physical sciences in the eighteenth and nineteenth centuries no easy generalizations can be made. Instead, we will consider three examples as representative of different modes of the relationship between religion and science.

Ruggero Boscovich, S.J. (1711–1787) epitomizes what John Haught has referred to as the "contact model" of relating science and religion, that is, looking for an open-ended conversation between the two. Contact "allows for interaction, dialogue, and mutual impact but forbids both conflation and segregation. It insists on preserving differences, but it also cherishes relationship" (Haught, 1995:18). Boscovich was born in Croatia and educated in Italy at the Jesuits' *Collegio Romano*, where he taught mathematics while studying for the priesthood. Ordained in 1744, he continued to research and publish widely in mathematics and astronomy. Among the first mathematicians on the Continent to adopt Newton's theories about gravitation, Boscovich correctly predicted that owing to the gravitational perturbations caused by Saturn and Jupiter, Halley's Comet would return late on its 1758 orbit around the sun. He is most famous for his *Theoria philosophiae naturalis* (1758), an attempt to offer a comprehensive explanation of the universe according to the principles of Newtonian mechanics and atomic theory. Boscovich envisioned not the hard extended atoms of Epicurus or Gassendi, but rather dimensionless points governed by attractive and repulsive forces that vary in proportion to distance, and in this he anticipated in some respects the idea of a unified field theory envisioned by Einstein. He remained a loyal Catholic even after the Jesuit order was dissolved in 1773, and it was in part because of his integrity and prominence as a scholar that Pope Benedict XIV was convinced to remove Copernicus' *De revolutionibus* from the Index of Forbidden Books.

Pierre Simon Laplace (1749–1827) presents a very different picture, reminiscent of the conflict models of Barbour or Haught, in which science swallows up or explains away religion (Barbour, 1997:78–82). Born in Normandy, Laplace abandoned his course of study for the priesthood at the age of twenty and threw himself into mathematics (Hahn, 2005:16–31). A brilliant theorist, he developed his nebular hypothesis of the formation of the solar system from the gravitational collapse and rotation of a cloud of dust and gas (Immanuel Kant had independently developed a similar theory in 1755). Laplace was an uncompromising theorist in contact with astronomers throughout Europe—he once tangled with Boskovich in 1776 over the proper measurement of cometary orbits—and published his comprehensive four-volume account of astronomy as *Mécanique céleste* (1799–1805). A thoroughgoing materialist, Laplace was convinced that determinism pervades the world, and he was contemptuous of mysteries such as the doctrine of transubstantiation (Hahn, 2005:222–224). In his private reflections, he was quite interested in how what he regarded as pre-critical myths had become transformed into Christian doctrines. He was strongly opposed by the deeply Catholic Italian mathematician Paolo Ruffini, who rejected Laplace's extension of probability theory to the moral sciences, on the grounds that it destroyed free will (Hahn, 2005:188).

In contrast to both Boskovich and Laplace stands French physical chemist Pierre Duhem (1861–1916), reflecting a "contrast model" of "sealing science and religion into separate containers" (Haught, 1995:15). A rigid ultra-Catholic, Duhem published extensively in the fields of both thermodynamics and the history of science. He occupied a difficult niche in the atmosphere of the French Third Republic, since maintaining allegiance to both his science and his Roman Catholicism required a delicate balancing act (Hiebert, 1986:438–440). From his compartmentalization of disciplines in his *"Physique de Croyant"* ("Physics of a Believer"), an essay appended to his *Physical Theory: its Object and Structure* (1914), we can infer that Duhem was a proponent of a two-languages approach to religion and science. Robert Deltete argues that Duhem developed a definite strategy in his defense of Christian belief (Deltete, 2004). First, in order to defuse the charge that his science was materialist and atheist, Duhem forwarded the "autonomy thesis," namely, that physics and metaphysics are fundamentally different enterprises, and that properly conducted, physics neither carries metaphysical implications nor requires metaphysical support. Second, Duhem used his "compartmentalist" position to defend the Catholic Church against the assaults of the positivist scientism then in favor with the Republicans. Third, he sought reciprocally to protect his science against fellow Catholics who wanted to use it for polemical purposes (Deltete, 2004). However, Duhem ultimately seems not to have succeeded in hermetically sealing off religion and science, arguing that the "teleological" development of the science of statics reveals divine providence, as it proceeds toward offering a true account of the way things are (Duhem, 1906).

CONCLUSION: A DIVERSITY OF APPROACHES IN COSMOLOGY AND RELIGION

The period from Nikolaus Copernicus to Pierre Duhem witnessed a revision in the Western understanding of cosmology and physics that was staggering in its dimensions and implications. Not only was the earth removed from its position at the dismal center of the cosmos (thereby dislocating Hell), but the doctrine of the perfection and immutability of the heavens was shattered. Orbits changed from being circular to elliptical, sunspots and lunar mountains revealed the corruptibility of celestial bodies, and the revised reckoning of the sun as simply one star among countless others relativized the position of earth-bound humanity almost beyond comprehension. Catholic thinkers played a central role in astronomy and cosmology in the Renaissance. The field shifted in early modernity, however, with unfortunate outcomes for both science and theology after the Galileo affair. The center of gravity of productive scientific investigation shifted toward England and other countries that were not hamstrung by the most authoritarian and bureaucratic features of the Counter-Reformation.

We conclude with a final note about the theological implications of the cosmological shifts explored in this chapter. As mentioned above, the pre-Copernican cosmology was integrated with theology to form an orderly scholastic synthesis of physics, astronomy, and theology. Each science in this hierarchy of disciplines operated from its own set of principles, and together they governed everything from the nature of matter and planetary motion to the geographic location of heaven and hell; the topography of Dante Alighieri's (1265–1321) *Divine Comedy* reflects the structure of the Aristotelian-Ptolemaic cosmos. And yet, inconsistencies were built into this synthesis, revealed in the recurrent question about whether everything in the celestial realm is more perfect than anything in the terrestrial realm. The stars may be physically closer to God, but they have no feeling as do creatures with animate souls, nor are they subject to the grace of salvation. Giovanni Battista Riccioli (1598–1671) concluded in his *Almagestum novum* (1671) that the earth with its living inhabitants is nobler than the sun (Grant, 2000:157–158).

Contrary to common belief, the Copernican shift to heliocentrism posed a theological threat not because it "dethroned" humanity, but rather because it elevated the earth and humanity out of the cosmic sump at the center of the cosmos (the "dung heap" as Pico della Mirandola referred to it in the fifteenth century) and toward the sphere of the Empyrean. Copernicus's student Rheticus wrote, "the globe of the earth has risen, while the Sun has descended to the center of universe" (Danielson, 2001:102). In a sense, then, Copernicus initiated the fateful and lengthy separation of theology from cosmology. It is hard to conceive of the Empyrean as the abode of God and the saints when the sun has now become simply a star like all the others twinkling in the night sky. The new cosmology was cold comfort to Blaise Pascal (1623–1662), who said that "the eternal silence of these infinite spaces frightens me" (Pascal, 2005:II.205). Catholics, like other Christians and the adherents of other religions, would have to come to terms with the implications of this new cosmology and integrate it into their worldviews. That is an unfinished project, part of the life story of each one of us.

Chapter 3

From the Garden of Eden to an Ancient Earth: Catholicism and the Life and Earth Sciences

INTRODUCTION

In the relationship between religion and science in the Christian West, the great stories of the sixteenth and seventeenth centuries involved fundamental questions about the composition and operation of the world. From Copernicus to Newton and beyond, the investigation of these questions led to a thorough revision of cosmology, physics, and chemistry. In the eighteenth and nineteenth centuries the great stories would be about the sciences of geology and natural history. The exploration of questions about why rock formations appear old and how species are related to one another would lead to a radically new understanding that the earth and life itself have historical dimensions. And just as Catholicism had slowly accommodated heliocentrism and an atomic theory of matter, Catholics, like other Christians, would adapt to a growing scientific consensus about the age of the planet and how life had come to diversify and flourish on it.

In June 1788, an amateur Scottish geologist named James Hutton (1726–1797) sailed with two colleagues along the coast of northeastern Scotland in search of rocks (Repcheck, 2003:12–24). He was hoping to find a particular formation that would substantiate his long-held theory that the earth was very old. Beaching their boat at Siccar Point, Hutton and his companions stared with awe at the cliff formation rising above the rocky beach. Horizontal strata of red sandstone overlay vertically oriented strata of grey micaceous schist, a classic "angular unconformity." For Hutton, who had studied geology for decades, this angular unconformity or juxtaposition of two different kinds of rock could mean only one thing. Clay alluvium had been deposited for eons along a coastline at the mouth of an ancient river, gradually hardening through metamorphosis into layered rock called schist. This stratum was tilted vertically by enormous forces deep within the earth and then sheared off by some other great force. The schist was later submerged again only to be covered by yet another layer of alluvium that hardened into red sandstone. Finally, the whole assemblage was once again raised above sea level. Hutton believed that all this geological change had been accomplished through forces acting uniformly throughout geological history. What it required above all was the vast expanse of time, orders of magnitude more time than the roughly 6,000 years of history provided by the biblical narrative. Hutton had found in the rocks testimony to his theory that the earth itself has an ancient history quite separate from that of humanity, and he published these ideas about "deep time" in his *Theory of the Earth*, 1888.

Nearly half a century later, a young Cambridge graduate named Charles Darwin (1809–1882) embarked upon a similar—if rather more extended—voyage of discovery. Finding medical studies distasteful, Darwin (on his father's insistence) had matriculated at Cambridge University to read theology in preparation for a clerical career. However, he was far more interested in studying natural history, and upon completing his degree he accepted an invitation to serve as expedition naturalist aboard the hydrographic survey vessel HMS *Beagle* that circumnavigated the globe from 1831 to 1836 (Browne, 1996:211ff; Desmond, 1994:101–191). Darwin spent his time cataloguing plant and animal specimens, collecting rocks and fossils, and conducting research into the natural history of the many lands he visited in South America and across the Pacific Ocean. He absorbed Hutton's ideas of "deep time" and uniformitarian geology through the pages of Charles Lyell's *Principles of Geology* (1830), which he had brought with him on board *The Beagle*. A keen observer of nature, Darwin developed a theory about the diversity of life and painstakingly assembled the supporting evidence. His explanation for the similarities between species—that they share common ancestry—and for the differences between them—that they have adapted to changing environments—was a challenge every bit as dramatic as that of Copernicus. Ironically, Darwin the student who so eagerly absorbed William Paley's providential interpretation of the adaptation of creatures to their environments would become Darwin the naturalist who stood Paley's argument for divine design on its head. This chapter recounts the story of how Catholic Christianity dealt with the discovery of the deep history of time and the supplanting of a miraculous with a naturalistic explanation for the diversity of life on earth. Not only science, but philosophy, theology, and every other discipline would be irrevocably altered.

Charles Darwin. (Image courtesy History of Science Collections, University of Oklahoma Libraries; copyright the Board of Regents of the University of Oklahoma)

NATURAL HISTORY BEFORE EVOLUTION

The idea that living things change over time was not born with Darwin but had a long period of gestation. Since antiquity students of nature had speculated about the origins of life, with Aristotle arguing that mice can be generated by decaying hay (as well as through normal procreation). For scholars in the Judeo-Christian tradition the creation story of Genesis 1–2 declared the starting point for terrestrial life, but it did not necessarily settle the question of whether God had created life in all its diversity in a mere six days. Was everything created at once, including dung beetles before there was dung, or maggots before there were decaying corpses? In *De genesi ad litteram* Augustine regarded a literal creation of everything in situ—including adult animals—as demeaning to the abilities of the Creator. While insisting that the universe had been brought into being in a single creative act, he thought that God had sown in nature the rational seeds (*rationes seminales*) of all that would later spring forth (McMullin, 1985:11–16). Augustine did not of course proffer an evolutionary theory; certainly he did not imply that one species comes from another (Woods, 1924:127–148). However, the concept of *rationes seminales* did introduce a developmental or gradualist dimension that became part of the medieval natural historical discussion.

In early modernity the unearthing of fossil sea shells on mountaintops and the discovery of distant cultures with chronological records older than the Genesis creation began to challenge the traditional biblical chronology (Rossi, 1984:123–151). But natural history was still bound by tradition, and the printed floral books of the early sixteenth century in some respects merely recapitulated the themes of medieval herbals. Likewise, Conrad Gesner's (1516–1565) handsomely printed *Historia Animalium* engaged the reader in the systematic study of animals, but also evidenced a credulity that a century later would be unacceptable: among the species to appear in Gesner and in Edward Topsell's *History of Foure-footed Beasts* (1608) were mythical creatures such as the manticora—part man and part lion!

By the late fifteenth century, however, some natural historians were beginning to examine the Roman natural historian Pliny (23–79 CE) with a decidedly critical eye (Findlen, 1996:58). The information conveyed by printing moved well beyond a crude iconographic tradition, and served to educate naturalists uniformly in far-flung parts of Europe (Eisenstein, 1979:543–574). Albrecht Dürer's detailed drawings of plants, for example, paid careful attention to their ecologies. The emblematic worldview—in which to know a creature was to know all of its literary associations—would gradually give way to the methodology championed by Francis Bacon (1561–1626), for whom personal observation was of central importance. Discovery of the divinely ordained laws of nature was a touchstone of the Royal Society (1660), echoing Federico Cesi's *Lyncaeorum Academia* in Rome. Pursuing sound and critical science was regarded as a deeply religious activity, and it is no accident that Nehemiah Grew's important empirical researches in botany found their highest expression in his *Cosmologia Sacra* (1701).

Natural historians continued to subordinate evidence to the judgment of theological assumptions, as did Nicholas Steno (1638–1686), a Danish Lutheran convert to Catholicism. A geologist ordained priest and then bishop, Steno made important discoveries about fossils and articulated three of the defining principles of the science of stratigraphy. However, the only framework available to him for interpreting his careful observations about marine fossils found on Italian mountains was Noah's flood, and his natural history remained firmly anchored within the received tradition of biblical theology (Cutler, 2003:115–122). Advances in scientific method and a critical understanding of observed phenomena fostered the empirical research of naturalists such as Robert Hooke and Anton van Leeuwenhoek, establishing the young science of biology on a firm footing by the end of the seventeenth century.

In the latter half of the seventeenth century the English "physico-theology" movement encouraged the closest partnership between religion and natural history that it would enjoy. Wrapping an ancient religious tradition in the mantle of the new science, most of its proponents maintained their Christian orthodoxy, although some (such as William Whiston) were much closer to the deist end of the spectrum of belief. The Boyle Lectures (founded in 1692) institutionalized this approach by providing a public forum for the articulation of the new science in the context of traditional Christian belief (Dahm, 1970:172). The Cambridge divine and scientist John Ray (1627–1705) perceived clear evidence of divine design in the intricate adaptations of plants and animals to their environments. He restated the design argument in a new guise in his treatise *The Wisdom of God Manifested in the Works of the Creation* (1691). But more importantly for biology, the book was a vast compendium of natural history, characterized by the observation of important nuances within species and the incipient recognition of ecological relationships. The most influential of these courses of lectures dealing with natural history in support of theistic belief was preached by William Derham (1657–1735) and published as *Physico-Theology, or, a Demonstration of the Being and Attributes of God from the Works of Creation* (1716).

The tradition of physico-theology depended fundamentally upon the cogency of the argument from design, or, in scholastic language, the proof from a "final cause." This teleological argument—found quite broadly within the Christian tradition—had been articulated in the context of systematic theology by St. Thomas Aquinas, who in turn had drawn upon St. John Damascene. But as the argument came to reflect and validate every imaginable natural historical detail as evidence of design, it began to wear thin. The case for design in nature as leading ineluctably to a creator met a forcible opponent in David Hume (1711–1776), a leading exponent of the Scottish enlightenment in the fields of history, economics, and philosophy. In his *Dialogues Concerning Natural Religion* (1750–1778) Hume attacked the argument from a number of angles, contending, for example, that order is perceived in mechanical processes such as the formation of snowflakes, not only where we would like to posit a designing mind. He also noted that the argument really only works where we compare an object we want to interpret as designed with another object we know to have been designed. We could only truly know that our universe had been designed if we were able to compare it with another instance of a universe known definitely to be a product of design. An even more damning critique (still bedeviling defenders of "Intelligent Design Creationism" today) is the problem of evil: if the universe has been designed by a wise creator, why has so much meaningless pain and suffering been designed or permitted to arise within the system?

Despite Hume's attack, physico-theology persevered with vigor and grace until Victorian times, nourished by the religious impulses of reverence and awe at the divine wisdom revealed in the works of creation. A classic example is the English clergyman Gilbert White's elegant collection of observations on local flora and fauna entitled *The Natural History and Antiquities of Selborne* (1789). A philosophically more rigorous contribution was William Paley's *Natural Theology* (1802), which interpreted the evidence of each species' special adaptation to its unique environment in terms of divine providential ordering. Paley's work embodying an anthropomorphic approach to natural history influenced a generation of students of impeccable Christian orthodoxy, including the young Charles Darwin. The eight Bridgewater Treatises of the 1830s epitomize the thematic depletion of the genre, although Charles Babbage's *Ninth Bridgewater Treatise* (1837) forwarded a modification of the thesis, namely, that an omnipotent God had the foresight to create laws permitting the development of new species under appropriate circumstances rather than through interference with the laws of nature themselves.

ENLIGHTENMENT AWAKENINGS TO THE HISTORY OF EARTH AND LIFE

The Professionalization of Natural History

Early in the eighteenth century natural history would split into two streams: one professional and increasingly secular; the other popular and persistently religious. Through a variety of channels cultural and intellectual factors would begin to erode the foundations of the physico-theological partnership of religion and science, to bring about its collapse in the nineteenth century. In one degree or another, the Enlightenment exaltation of reason at the expense of revelation is implicated in most of these factors, but social factors such as strenuous competition from academic natural historians also played an important role.

Swedish botanist and zoologist Carl Linnaeus (1707–1788) regarded himself as recorder of God's creation in creating his system of taxonomic nomenclature, but in his work he avoided as consistently as possible any appeal to God for causal explanations of natural historical phenomena. After 1700 the increasing sophistication of research tools and instrumentation, and the establishment of endowed chairs in European universities, led inevitably to the professionalization of the field. Natural history was on its way to becoming the largely secular professional discipline of biology that it would be in the nineteenth century, and there was less and less scope for amateur collectors of specimens to make respectable contributions. Impelled by the research of figures such as Linnaeus in

Sweden, Georges-Louis Leclerc, Comte de Buffon (1707–1788) in France, and Albrecht von Haller (1708–1777) in Germany, natural history on the professional level discarded in theory (if not in fact) the religious assumptions of physico-theology. As scientific sophistication spread to the wider culture, discoveries that had provoked awe and reverence in the early physico-theologians became regarded as merely commonplace.

From Teleology to Naturalism

An important Aristotelian legacy was the notion of *telos* signifying "end" or "purpose." For 2,000 years teleology had served as a fundamental organizing principle in science, and it played an especially crucial role in natural history. Organisms were believed to develop according to a preconceived plan and organs were regarded as having been designed to serve specific purposes, just as animals were believed to have been located in particular habitats to serve the needs of humans. In the eighteenth century this basic principle began to falter, and while it would be anachronistic to suggest that by 1859 teleological thinking already lay shipwrecked on the shoals of naturalism, its piecemeal dismantling cannot be ignored. Despite their personal and methodological differences as scientists, Linnaeus, Buffon, and Haller shared basic assumptions about the existence of final causes and immutable plans regarding the objects of their study. In contrast, their successors in the next generation of natural historians uniformly relied upon the assumptions of Enlightenment science, discarding as useless tools the teleologies and immutable plans presupposed from Aristotle to John Ray and William Derham. Their intentionally non-teleological approach found philosophical legitimation in Kant's *Critique of Teleological Judgment* (1790).

Contributing to the dissociation of natural history from a religious or spiritual interpretation of nature was an extension of the seventeenth-century mechanistic cosmological model into the biological sphere. The reintroduction of a Lucretian atomic theory of matter, purged of its atheistic elements, had reduced physical reality to matter in motion under the influence of Galileo and Descartes, and already essentially excluded the vast continuum of "vital" powers intrinsic to the Aristotelian universe. The logical sequel was to extend this reductionism to life itself, and to provide a purely naturalistic explanation of life from a mechanico-chemical perspective (Roger, 1986:279–286). The mechanistic interpretation of life may not ultimately have won many converts, but it further eroded the supports for a religious interpretation of nature.

"Higher Criticism" of the Bible

In the eighteenth century scholars began to recognize that the Book of Genesis is not a reliable guide to the history of the earth. There were several dimensions to this. The interest of Renaissance humanists in reading the Bible in the original languages of Hebrew and Greek had initiated a process of critical inquiry that gathered momentum in succeeding centuries. In 1773 French physician Jean Astruc published a book entitled *Conjectures on the original documents that Moses appears to have used in composing the Book of Genesis*. (A convert to Catholicism, Astruc was the son of a Protestant minister from a family with medieval Jewish origins.) Remarking upon the fact that the biblical author alternated the use of the two Hebrew names for God in Genesis, *Elohim* and *Jehovah*, he proposed that Moses had amalgamated two preexisting documents. Astruc's insight was adopted by the Protestant theologian Johann Gottfried Eichhorn (1752–1827), the "father of modern biblical criticism." Independently, a Scottish Catholic priest, Alexander Geddes also anticipated the German school of higher criticism (Fuller, 1984:60–82). This historical-critical movement—which recognized the multiple authorship of the Hebrew Bible and that much of its content was derived from the oral or written traditions of other Semitic cultures of the Ancient Near East—began to call into question the reliability of the biblical witness to secular history, including the stories of creation (*Genesis* 1–2), and the account of a universal flood (*Genesis* 6–9).

The Historicization of Earth and Life Sciences

The Book of Genesis does not indicate the date on which the world was created, an omission that not a few scholars had tried to rectify (McCalla, 2006:28–39). The most famous calculation was made by the same biblical exegete who made the first serious examination of the Septuagint (the Greek translation of the Hebrew Bible): James Ussher, Archbishop of Armagh and Anglican Primate of All Ireland. In his scholarly *Annals of the Old Testament, Deduced from the First Origins of the World*, Ussher reckoned creation to have occurred at nightfall preceding October 23, 4004 BC (Roberts, 2007:41). Ironically, within a century the calculation of creation—in both method and rationale—would be an obsolete exercise. By the mid-eighteenth century the evidence from literary criticism militating against every passage of the bible being literally true coincided with another serious challenge to the received tradition of natural history: a secularizing trend toward the "historicization" of the life and earth sciences.

Two growing mountains of evidence—one geographical and the other temporal—suggested that the biblical cosmogony and global flood story were seriously flawed and becoming untenable (Cohn, 73–120). First, almost from the moment of the European discovery of the "new world," an endless stream of information about previously unknown plants and animals began inundating the consciousness of natural historians. Whereas

John Ray listed 1,500 species of animals in 1691, Carl Linnaeus in the 1758 edition of his *Systema Naturae* included 4,400 species of quadrupeds alone, and further geographical exploration offered no end to this explosion of knowledge. It began to appear impossible to reconcile such an abundance of species with a hexaemeral (six-day) creation. Moreover, the fact that marsupials are found only in Australia and that Africa and the Americas have their own collections of indigenous animals rendered implausible at best the story of Noah having saved at least two of each species in his ark. Nor could the Noah story credibly account for the postdiluvial radiation of all species throughout the world from Mount Ararat (Browne, 1983:1–31). Natural historians henceforth would increasingly be forced into the uncomfortable position of having to choose between their empirical evidence and the dictates of a literalist biblical tradition.

The temporal factor influencing the historicization of natural history was the gradual discovery of the "deep history" of time (Rudwick, 2007). As noted above, James Hutton concluded that the sedimentary strata of the earth and the fossils they contained were far older than the few thousand years permitted by a literal reading of Scripture. His belief in the antiquity of the earth was supported by the French vertebrate paleontologist Georges Cuvier (1769–1832), who established as fact that species had in the past been driven extinct. In his *Discourse on the Revolutionary Upheavals on the Surface of the Earth* (1812), Cuvier agreed that the earth is immensely old. In contrast to Hutton's "uniformitarian" assumption, however, Cuvier took a contrary position that later was labeled "catastrophism"—that the earth has periodically undergone major regional upheavals triggering spasms of extinction. Cuvier preferred to avoid supernatural connotations by applying the term "revolutions" to what he thought were purely natural events. "Catastrophists" such as the Reverend William Buckland in England, however, sought to unite scriptural and geological history, arguing that the most recent revolution had been the Biblical Flood (Brooke, 1991:192–225; UCMP Web page).

Catholic participation in the geological debates was delayed, with the first contribution being French Abbé Sorignet's *Sacred Cosmogony* (1854; trans. 1862), in which he declared that geology that is not in harmony with scripture is unsound. An American Catholic priest, Clarence Walworth, rejected Sorignet's argument in his *The Gentle Skeptic* (1863), contending that the six days of creation were merely metaphorical expressions employed by the author to classify the works of creation (O'Leary, 2006:11). Among the more comprehensive contributions to the discussion was Irish priest Gerald Molloy's *Geology and Revelation* (1870), which relied substantially upon Charles Lyell's *Principles of Geology* (O'Leary, 2006:12). He also drew upon *Twelve Lectures on the Connection between Science and Revealed Religion* (1836) by Cardinal Nicholas Patrick Wiseman (later made first cardinal archbishop of Westminster). Like Walworth, Molloy was concerned about maintaining the credibility of Catholics in intellectual discussions, and emphasized the "two books" metaphor, the medieval and early modern trope that God is the author of both the "Book of Nature and the Book of Scripture" (Hess, 2003:124–126). Through a range of geological arguments Molloy's recurrent theme was the "accommodationism" employed by Galileo: the Bible was written not to teach science, but to serve the spiritual needs of believers, and the scriptural authors accommodated the needs of simple people when speaking about scientific matters.

The fascinating history of evolutionary ideas before Darwin is as convoluted as that of the geological discovery of deep time, and too intricate to review here (see Bowler, 2003:48–140). Suffice it to say that Darwin matured as a naturalist in a context in which evolutionary thinking had been current for half a century. His grandfather Erasmus Darwin (1731–1802) was a physician and physiologist who proposed a theory of organic development in *Zoonomia* (1796). As a deist, Erasmus Darwin argued that in order to meet the challenges posed by their external environments God had designed living organisms to be self-improving. Similar ideas were proposed in France by Cuvier's rival, the strict materialist Jean Baptiste Lamarck (1744–1829), a professional naturalist. In his *Philosophie Zoologique* (1809), Lamarck developed a theory of linear progress (in contrast to Erasmus Darwin's theory of divergence) and of evolution through the transmission of acquired characteristics. Though he was largely discredited for this, he was supported by Paris zoology professor Étienne Geoffroy Saint-Hilaire. Lamarck made the first significant attempt to construct a theory of descent of all living beings from common ancestry (Bowler, 2003:85–94). In this atmosphere, William Paley's rearticulation of the argument in *Natural Theology* (1802) for a divinely designed and providentially arranged world was scientifically already out of date.

Moreover, the tradition of constructing "chronicles of the world" underwent a gradual shift from portrayals of human history as synchronous with that of the world to a variety of schemata that included increasingly greater lengths of prehuman earth history (Rudwick, 1986:296–321). In the previous hundred years the timeline of earth history had lengthened dramatically from the 6,000 years of Archbishop Ussher's calculation to the hundreds of thousands of years of German geologist C. G. Füchsel (1762), a defender of the Mosaic scriptural account who had interpreted the six days of creation as long ages. A decade later James Hutton mused that in the record of the terrestrial crust "we find no vestige of a beginning, no prospect of an end," and in 1787 his contemporary Abraham Werner suggested an age of a million years. Charles Darwin judged that it would require hundreds of millions of years for evolutionary change to happen, and Lord Kelvin in 1862—basing his reckoning upon the cooling of the sun—calculated between 24 and 400 million years.

In any case, by the mid-nineteenth century the earth was recognized by scientists as being tens of millions of years old and organic life as in some fundamental way developmental. Professional natural historians had every good reason to believe that the history of creation was not coterminous with human history, and that over time species have come into existence and gone extinct. A permanent rift developed between scriptural geologists who insisted that Genesis interprets nature, and secular geologists and biblical exegetes of the historical-critical school who maintained that nature interprets Genesis (Moore, 1986:335–340). At the same time progressive scripture scholars appreciated the Bible as a complex, multisourced document, in which the hexaemeral creation story in Genesis served the function of cosmogonic myth, and the Noachian deluge was an archetypal story that need not be read literally.

When in this intellectual cultural context Charles Darwin developed a carefully substantiated case for evolution in *On the Origin of Species* (1859), his theory met a mixed reception, with some theologians accepting it and some scientists opposing it (Dupree, 1986:355–362). Darwin himself gradually abandoned Christianity as he found its teleological presuppositions to be incompatible with empirical evidence. John Brooke has inferred that Darwin's gradual loss of traditional faith had as much to do with his emotional response to the tragic death of his daughter Annie as it did with his developing perspective on natural selection (Brooke, 2002:38–41). The theory of evolution was in some respects consonant with his agnosticism, but not necessarily causative of it.

CARDINAL NEWMAN: SCIENCE IN CATHOLIC THOUGHT AT MID-CENTURY

In order to gauge the relationship between science and religion in mid-nineteenth-century Catholic perspective, let us consider the thought of one of the greatest of English Victorian thinkers, John Henry Newman (1801–1890). An Anglican priest and member of the High Church Oxford Movement, Newman converted to Catholicism in midlife, and later was created cardinal. In both phases of his religious life Newman was strongly motivated by pastoral concerns. Not long before the appearance of *The Origin of Species*, he wrote *The Idea of a University* (1858) offering some intriguing Catholic perspectives on the proper relationship between theology and science. Or rather, we should say that he offered a perspective that allowed interpretation in at least two ways. First, Newman articulated what is essentially a two-languages position, arguing that since theology is the "philosophy of the supernatural world" and science the "philosophy

John Henry Newman (1801–1890). An Anglican priest and member of the High Church Oxford Movement, Newman converted to Catholicism and was created cardinal. In his classic *The Idea of a University* (1858) he argued for the complementarity of theology and science. (Artist: Herbert Rose Barraud. Credit: National Portrait Gallery, London.)

of the natural, Theology and Science, whether in their respective ideas, or again in their own actual fields, on the whole, are incommunicable, incapable of collision, and needing, at most to be connected, never to be reconciled" (Newman, 1907:431).

But, second, Newman remained serenely confident that truth is ultimately one, even if it is known through different modalities. The theologian has no reason to become alarmed that discoveries arrived at by means of any scientific method other than theology can truly contradict formal religious dogma. In fact, Newman argued, theology is the one science that from its "sovereign and unassailable position" can remain epistemologically unperturbed:

He is sure, and nothing shall make him doubt, that, if anything seems to be proved by astronomer, or geologist, or chronologist, or antiquarian, or ethnologist, in contradiction to the dogmas of faith, that point will eventually turn out, first, *not to be proved*, or, secondly, not *contradictory*, or thirdly, not contradictory to any thing *really revealed*, but to something which has been confused with revelation. (Newman, 1907:466–467)

Newman cautioned both scientists and theologians to observe the integrity of their respective disciplines and to avoid trespassing on intellectual territory outside their competence. As a nontheologian Galileo would have been better off, he thought, "to hold his doctrine of the motion of the earth as a scientific conclusion" rather than pontificating in the *Letter to the Grand duchess Christina* about what the scriptures teach us. Speaking of contemporary theologians, Newman scolded

religious men, who, from a nervous impatience lest Scripture should for one moment seem inconsistent with the results of some speculation of the hour, are ever proposing geological or ethnological comments upon it, which they have to alter or obliterate before the ink is well dry, from changes in the progressive science, which they have so officiously brought to its aid. (Newman, 1907:472)

This position clearly reflects a commitment to theology in dialogue with scientific culture, not a theology hermetically sealed off from science. Newman seems neither to have specifically endorsed or denied human evolution, but he was among the first theologians to accept the scientific principle of the mutation of species. (Elder, 1996:75–76). As a friend of the English biologist St. George Mivart, Newman endorsed the latter's evolutionary ideas, speculating in a letter to E.B. Pusey in 1870 that Darwin's views about human descent did not contradict scripture.

EARLY VATICAN REACTIONS TO DARWINISM

We should not expect to find a monolithic Catholic reaction to evolution, on the part either of the papacy, or of the bishops and priests, or of Catholic scholars or lay persons. Turning centuries of established tradition on its head was no mean task, as the biblical critics who contributed to the Anglican *Essays and Reviews* (1860) were well aware. Various forces within Catholicism would serve to modify or deflect the trajectory of the gradual acceptance of evolution. The earliest official statement by the Catholic hierarchy based its rejection of human evolution on both scripture and tradition. In 1860, at a provincial Council of Cologne, during deliberations about principles of Christian anthropology, the council declared as contrary to scripture and faith both the opinion that the human body was derived from the spontaneous transformation of an inferior nature, and the view that the entire human race had not descended from Adam (Alszeghy, 1967:25). However, it is important to note that the primary objection was to "spontaneous" transformation, permitting the hypothesis that God might have aided in the evolutionary process.

The reaction of the papacy to evolution was strongly conditioned by political circumstances. In the wake of the French Revolution (1789–1799), the Catholic Church had been split between liberal and conservative factions, with more bishops supporting the ultramontane or conservative party. Ultramontanists found ideas about civil equality, religious toleration, freedom of the press, and the separation of church and state threatening. Not surprisingly papal policy for most of the middle of the century reflected rigidity:

An institution so politically, economically, and socially conservative was not likely to be progressive, flexible, and innovative when responding to intellectual challenges in religious matters. (O'Leary, 2006:45–46)

In the face of such a wide range of challenges, Pope Pius IX issued in 1864 his *Syllabus of Errors*, a broad attack on liberalism in politics, the sciences, and critical biblical scholarship. The *Syllabus* not only reaffirmed the supremacy of theology over philosophy, but sanctioned a particular methodology by condemning the proposition that "The method and principles by which the old scholastic doctors cultivated theology are no longer suitable to the demands of our times and to the progress of the sciences" (*Syllabus*, 13). Pius insisted that philosophy should operate with one eye on supernatural revelation. Even taking into account the Vatican's justifiable need to defend itself against the aggressive nature of some anti-Catholic factions within Italy, the *Syllabus of Errors* comes across as stridently defensive. In the year of its publication (1864) Darwin's *Origin of Species* was translated into Italian, initiating rancorous debate that the Church interpreted as an attack on its teachings (O'Leary, 2006:52).

A few years later Pope Pius IX convened the first Vatican Council (1868) to reassert the power of the papacy. His boldest move was to push through the College of Cardinals the doctrine of papal infallibility on doctrinal matters of faith and morals when the pope is speaking *ex cathedra* (that is, "from the chair" of the Roman Pontiff). The delegates to the council also argued in the *Dogmatic Constitution on the Catholic Faith* to defend theology against the false claims of philosophy and science:

Hence all faithful Christians are forbidden to defend as the legitimate conclusions of science those opinions which are known to be contrary to the doctrine of faith, particularly if they have been condemned by the Church; and furthermore they are absolutely bound to hold them to be errors which wear the deceptive appearance of truth. (Pope Pius IX, *DCCF*, 1870a:4.9)

But the bishops did not explicitly name developmental biology, and the declaration that "faith and reason [can] never be at odds with one another but mutually support each other" leaves considerable room for interpretation. The *Dogmatic Constitution on the Church* did not forbid the sciences to employ their own proper principles and method, although it did place constraints upon them:

While she admits this just freedom, she takes particular care that they do not become infected with errors by conflicting with divine teaching, or, by going beyond their proper limits, intrude upon what belongs to faith and engender confusion. (Pope Pius IX, *DCC*, 1870b:10)

Pope Leo XIII declared in *Providentissimus Deus* (1893) that the science is "admirably adapted to show forth the glory of the Great Creator, provided it be taught as it should be," but that if it be "perversely imparted to the youthful intelligence" it can prove most fatal by destroying the principles of true philosophy and corrupting morality. Science thus serves a protective role, assisting in detecting attacks on the Sacred Books, and in refuting them. In fact,

There can never, indeed, be any real discrepancy between the theologian and the physicist, as long as each confines himself within his own lines, and both are careful, as St. Augustine warns us, "not to make rash assertions, or to assert what is not known as known." (Leo XIII, 1893:13)

In fact, Pope Leo appealed to the rule of St. Augustine that in cases of conflict, it was the Church's right and responsibility to enforce an interpretation of scientific evidence consonant with scripture:

Whatever they can really demonstrate to be true of physical nature, we must show to be capable of reconciliation with our Scriptures; and whatever they assert in their treatises which is contrary to these Scriptures of ours, that is to Catholic faith, we must either prove it as well as we can to be entirely false, or at all events we must, without the smallest hesitation, believe it to be false. (Leo XIII, 1893:13)

But even here it seems to me we see an ambiguity pregnant with possibility. Almost in echo of the language used in the Galileo controversy—that the scriptures teach us how to go to heaven not how the heavens go—Pope Leo XIII declared that we must bear in mind that the sacred writers, "or to speak more accurately, the Holy Ghost Who spoke by them, did not intend to teach men these things (that is to say, the essential nature of the things of the visible universe), things in no way profitable unto salvation." In other words, biblical writers were not writing as scientists in quest of the secrets of nature, but rather described and dealt with their subject in figurative language, or in terms that were commonly used at the time. Even the "Angelic Doctor," St. Thomas Aquinas, maintained that the sacred writers "put down what God, speaking to men, signified, in the way men could understand and were accustomed to." Pius XII would articulate the same idea in *Divino Afflante Spiritu* (1943).

In Scripture divine things are presented to us in the manner which is in common use amongst men. For as the substantial Word of God became like to men in all things, "except sin," so the words of God, expressed in human language, are made like to human speech in every respect, except error. (Pius XII, 1943:37)

The key issue for Catholics (as for others) was whether we are bound to reinterpret the findings of science so as to be consonant with a theological worldview that in essential respects was out of date. Still, evolution as a Catholic problem had less to do with biblical literalism than it did with the Church's adherence to Neo-Scholasticism. When the latter began to fade as a principal metaphysical system, the way was opened for an exploration of evolutionary ideas.

ST. GEORGE JACKSON MIVART: EVOLUTION IN THEISTIC PERSPECTIVE

One of the early significant endorsements of the theory of evolution from a solidly theistic perspective was by the English anatomist and convert to Catholicism, St. George Jackson Mivart (1827–1900). In his *On the Genesis of Species* (1871), Mivart expounded arguments important enough for Darwin to pay serious attention to in subsequent editions of on the *Origin of Species*. Mivart had taught himself zoology, and developed a specialization in the anatomy of newts and monkeys. With the sponsorship of Thomas Huxley, he became a Fellow of the Royal Society and made the acquaintance of Darwin. The relationship soured after Mivart's publication of his own book and worsened following the appearance of Darwin's *Descent of Man*, not so much on account of Mivart's opposition to elements of Darwin's theory as because of the personal nature of his attacks upon the latter. Mivart was a strong believer in biological evolution with the important stipulation that it should be framed in opposition to an atheistic worldview. As a response to what he feared was the challenge of a thoroughly naturalistic evolution, he placed a protective belt around the creation of the human soul in order to integrate biological evolution with the Roman Catholic doctrine of theological anthropology.

Mivart supported his arguments for the compatibility of the theory of evolution with Christian doctrine by adducing the opinions of the early church fathers and medieval theologians as to how the universe might be regarded as unfolding from "seeds." Most important among these was Augustine's "derivative sense in which God's creation of organic forms is to be understood," namely the conferral on the world of the power to evolve from *logikoi spermatikoi*, or "word empowered seeds" (Mivart, 1871:281). As a preliminary line of argument Mivart declared that creation in fact forms part of revelation, and that revelation appeals for its acceptance to reasons. Thus we are led in the direction of a philosophical theology in which the intelligent acceptance of theism on purely rational grounds served to prepare the way for acceptance of the reasonableness of revelation (Mivart, 1871:261). Part of Mivart's strategy in the early part of his twelfth chapter, on "Theology and Evolution," was to claim that Darwin and Spencer had created straw person characterizations of teleology or "ultimate cause." Next he argued that the claims of theism are made entirely in analogical language (Mivart, 1871:264), and that since theists declare at the outset their conception of God to be utterly inadequate, scientists who ridicule belief in God as Creator on the grounds of lack of physical evidence are guilty of what we today might call "scientism" (Mivart, 1871:272–273).

Mivart constructed protective belts at two points to safeguard Christian doctrine from physicalist reductionism. The first is "absolute or primary creation," which stands completely outside the purview of physical science. While derivative creation—or the action of evolution upon the animate world to provide a diversity of species—is not a supernatural act, primary creation is a supernatural act. Mivart believed that careful distinction between these categorically different levels of creation not only could dispel theological reluctance to adept evolution, but could obviate a simplistic rejection of the doctrine of creation on the part of scientists. The second protective belt was drawn around the human soul in a move often repeated by Catholic theologians up to Popes John Paul II and Benedict XVI. Mivart could in no way countenance the derivation of the spiritual component of humanity from the slime of the earth.

One of the least satisfactory aspects of Mivart's argument for the integration of evolutionary theory with a perspective of providentialist theology was his treatment of the problem of pain suffered through eons of evolution. He tried to minimize animal pain, by suggesting that

only during consciousness does it exist, and only in the most highly organized men does it reach its acme. The author has been assured that lower races of men appear less keenly sensitive to physical pain than do more cultivated and refined human beings. Thus only in man can there really be any intense degree of suffering, because only in him is there that intellectual recollection of past moments and that anticipation of future ones, which constitute in great part the bitterness of suffering. (Mivart, 1871:277)

Today we would regard this as a cavalier dismissal of the reality of non-human animal experience (to say nothing about its pseudo-distinction between races of humans). Whatever truth there may be to the differentiation between physical and psychological pain, Mivart's minimization of the former was not a sufficiently convincing refutation of Darwin's recognition that the process of natural selection has involved countless generations of conscious beings in a systematic web of suffering. Mivart's metaphysic of design predisposed him to an inevitably beneficial reading of evolution: "The natural universe has resulted in the development of an unmistakable harmony and beauty, and in a decided preponderance of good and happiness over their opposites" (Mivart, 1871:278). This reads uncomfortably and unconvincingly like a utilitarian justification of suffering, or like Paley's imagined world in *Natural Theology* of his vicarage garden buzzing with joyful insects. Later thinkers (including Catholics) began to reject such simplistic harmonization, realizing that no Christian theology of nature is adequate that does not at least address the theodicy problem seriously (Haught, 2005:12).

When Mivart taught that God could infuse a soul into a body prepared by a preceding process of evolution, he was strongly criticized by some Catholic scholars, but the authorities did not intervene; in fact, in 1876 Pope Pius IX conferred upon him the degree of doctor of philosophy. In 1891, Cardinal Zeferino Gonzáles corrected Mivart's theory that God could render the hominid body capable of receiving a spiritual soul through the "special action" of evolution. Gonzáles thought this correction was necessary to preserve human dignity and to distinguish us from the beasts (Garrigan, 1967:84).

What landed Mivart in trouble with Rome had nothing to do with evolution, but rather with his questionable ideas about the theology of hell. Attempting to mitigate for apologetic reasons the traditional gloomy doctrine of eternal punishment, he argued in 1893 that most human beings are incapable of committing mortal sin, and that after death, influenced by their natural love for God they will attain a state of mortal happiness commensurate with their abilities. Mivart's articles on this subject were placed on the Index in 1893, but beyond the official pronouncement, there was no substantial campaign of enforcement or repression. It was Mivart himself who rashly insisted on provoking the authorities, who ultimately prohibited him from receiving the sacraments (Artigas, 2006: 248–264).

EVOLUTION AND RELIGION: GERMANY AND FRANCE

Dupree has sketched the diversity of nineteenth-century Protestant reaction to Darwin's theory of evolution, showing that there were scientists and churchmen both in support of and opposed to the theory (Dupree, 1986:355–362). Naturally we might expect to discover a comparable diversity among Catholic theologians and scientists. In European circles the fortunes of evolution varied widely, with scholars who had significant experience in biological science taking a more open and exploratory position than their more tradition-minded colleagues. The Jesuit German biblical exegete J. Knabenbauer offered the following careful judgment in 1877, published in *Stimmen aus Maria-Laach* ("Voices from Maria-Laach"):

Considered in connection with the entire account of creation, the words of Genesis cited above proximately maintain nothing else than that the earth with all that it contains and bears, together with the plant and animal kingdoms, has not produced itself nor is the work of chance; but owes its existence to the power of God. However, in what particular manner the plant and animal kingdoms received their existence: whether all species were created simultaneously or only a few which were destined to give life to others: whether only one fruitful seed was placed on mother earth, which under the influence of natural causes developed into the first plants, and another infused into the waters gave birth to the first animals—all this the Book of Genesis leaves to our own investigation and to the revelations of science, if indeed science is able at all to give a final and unquestionable decision. In other words, the article of faith contained in Genesis remains firm and intact even if one explains the manner in which the different species originated according to the principle of the theory of evolution. (Knabenbauer, 1877:78)

What is particularly intriguing about this favorable judgment on the evolutionary hypothesis is that Knabenbauer's order—the Society of Jesus—were staunch opponents of the anti-Catholic secularizing forces in Bismarck's *Kulturkampf*, or "culture war." Whether Knabenbauer's primary purpose was to defend the integrity of science or the credibility of Catholicism, he demonstrated that one could be a loyal son of the Church and at the same time integrate evolutionary thinking into theology.

Darwinism fared less well in nineteenth-century France, where Lamarckian ideas continued to flourish, and where Catholicism, in a fragile political position since the French Revolution, was preoccupied with its own survival amidst the welter of secular forces and occultist elements. French Catholicism at the turn of the century tended to exalt the mysterious and miraculous (and even the irrational), and did not offer fertile ground for an assimilation of Darwinist evolutionary ideas (Burrow, 2000:51, 225–226).

Among the important exceptions to this generalization was Fr. Dalmace Leroy, a French Dominican whose lifelong fascination with natural history led him in 1887 to publish *The Evolution of Organic Species*. In response to reviews critical of his book he published a revised and expanded edition in 1891, under the altered title *Evolution Limited to Organic Species*, explaining that he had carefully excluded Adam and Eve from consideration in the evolutionary story. Nevertheless, his book was denounced to the Index in 1894, and following lengthy negotiations Leroy agreed to retract it in 1895. He retained reservations, however, for he sincerely believed that the credibility of the Catholic Church was at stake in its steadfast refusal even to consider the evolutionary preparation of the human body for reception of the infused soul. The cardinals forbade Leroy's book but did not publish the decree condemning it. The detailed records in the Vatican Archives show that this was by no means a simple affair, and that there was incertitude among the theologians of the Congregation of the Index. The case stands as clear evidence that the Catholic Church in the generations after Darwin had no official doctrine regarding evolution (Artigas et al., 2006:52–123).

ENTHUSIASTIC APPROPRIATION: THE CASE OF JOHN AUGUSTINE ZAHM

The vigorous discussion of the theory of evolution at the turn of the century reflects the struggle internal to the American Catholic community over their integration into American culture. Free scientific inquiry and the separation of church and state were values not to be taken lightly. Appleby has shown that the reception of Darwinism was conditioned in part by the ongoing "conflict between 'scholastic' (i.e., Aristotelian) and 'modern' (i.e., Newtonian) methods and ways of conceptualizing nature and human development" (Appleby, 1999:198). Evolution was therefore judged according to the norms of the Neoscholastic worldview that had become entrenched in Catholic seminaries in the nineteenth century.

John Zahm (1851–1921) was one of the foremost American apologists for the harmonization of evolutionary theory with Catholic dogma. Ordained a priest in the Holy Cross order in 1875, he was a professor of physics and chemistry at the University of Notre Dame. In the year of publication of his book *Evolution and Dogma* Zahm was transferred to Rome as procurator general for the Holy Cross Community, which some opponents incorrectly interpreted as censure for his dubious views. This was in fact not the case, but the appearance of the French and Italian editions of his book inevitably led to intensified debate about these issues for a year, until the Sacred Congregation of the Index issued an injunction against its further publication

John Augustine Zahm (1851–1921). Holy Cross priest and professor of physics and chemistry at the University of Notre Dame, who become one of the foremost American apologists for the harmonization of evolutionary theory with Catholic dogma. (Courtesy of the Notre Dame Archives)

and distribution in 1899, although this injunction appears never to have been enforced.

Zahm demonstrated an ease and familiarity with evolution that sounds surprisingly contemporary to us. He noted that Darwinism "is not evolution, as is so often imagined, but only one of numerous attempts which have been made to explain the *modus operandi* of evolution" (Zahm, 1978:207). He was well aware of the baggage Darwinian evolution bore in its being associated with atheism, citing Canadian geologist William Dawson who maintained that "the doctrine of evolution carried out to its logical consequences excludes creation and Theism" (Zahm, 1978:209). He was also alert to evolutionary controversies in Europe, appealing, for example, to the opinion of French priest Monsabré that

far from compromising the orthodox belief in the creative action of God [evolution] reduces this action to a small number of transcendent acts, more in conformity with the unity of the Divine plan and the infinite wisdom of the Almighty, who knows how to employ secondary causes to attain his ends. (Zahm, 1978:212)

Zahm recognized the paucity of fossil transitional forms, and appealed to Darwin's own treatment of this regarding our incomplete knowledge of the geological record (Zahm, 1978:161, 173). He was convinced that although we do not at present understand the production of variation—Mendelian genetics would not be incorporated into the neo-Darwinian synthesis for some time yet—we would eventually arrive at this understanding. And he critically reviewed the evidence on both sides of the controversy about Lamarckian transmission of acquired characteristics, weighed Darwinian and non-Darwinian approaches, and concluded that a true, all-embracing theory of evolution still awaited scientists (Zahm, 1978:195–202). Zahm retained serene confidence that progressive science and revealed theology are consonant endeavors:

Whatever may be the final proved verdict of science in respect of man's body, it cannot be at variance with Catholic dogma. Granting that future researches in paleontology, anthropology, and biology, shall demonstrate beyond doubt that man is genetically related to the inferior animals, and we have seen how far scientists are from such a demonstration, there will not be, even in such an improbable event, the slightest ground for imagining that then, at last, the conclusions of science are hopelessly at variance with the declarations of the sacred text, or the authorized teachings of the Church of Christ... We should be obliged to revise the interpretation that has usually been given to the words of scripture which refer to the formation of Adam's body, and read these words in the sense which evolution demands, a sense which, as we have seen, may be attributed to the words of the inspired record, without either distorting the meaning of terms or in any way doing violence to the text. (Zahm, 1978:364–365)

Zahm's book invited attention in Rome in part because of the existing campaign against "Americanism." New World political values were often regarded with suspicion by conservative Europeans, for American Catholics who had adopted the values of freedom of the press, liberty of conscience, and the spirit of free scientific inquiry were less likely to follow Vatican dictates. Nevertheless, Zahm was deferential to authority, and when rumors reached him that his book was about to be placed on the Index, he promptly wrote to the publisher of the Italian edition to slow its distribution. The decree of condemnation was never published, and Zahm never issued a retraction. Convinced that he had worked for the honor and glory of God in writing about evolution and dogma, he was content to follow the orders of the hierarchical church he loved and served (Artigas, 2006:194–196).

Artigas, Glick, and Martinez show that the popular assumption of Rome having taken a heavy-handed role in quelling Catholic enthusiasm for biological evolution is quite simply false. The Jesuit journal *La Civiltà Cattolica*—whose editors enjoyed a close relationship with the Vatican—launched an extended campaign against the theory, but their influence was limited. None of Charles Darwin's books were placed on the Index, nor were those of Thomas Huxley ("Darwin's bulldog"), Herbert Spencer, or Ernst Haeckel. The six cases that came to the attention of the Congregation of the Index during the quarter century under review all involved books written by Catholics who had attracted ecclesiastical attention, presumably because their works had greater potential to disturb the life of the Church (Artigas, 2006:14). Only the case of the Florentine priest Rafaello Caverni reached the ultimate conclusion of condemnation. Even here, however, the reason for the condemnation of his *New Studies of Philosophy* (1877) was not made public (Caverni was careful to except human beings from his defense of the theory) and the word "evolution" does not appear in the title (Artigas, 2006:35, 50–51).

That there was no Catholic conspiracy against evolution is further clarified by the fact that the Vatican investigations of proponents of evolution were undertaken not by its doctrinal authority—the Holy Office of the Inquisition—but rather by the congregation of the Index. Much had changed since the days of the Galileo affair, an event that seems to have been present in the consciousness of the participants on both sides. In 1616 and 1632 Galileo's ideas were condemned quite quickly and without much time for deliberation. In contrast, the discussion of evolution had lasted for several decades, but despite established theological opposition, these cases led to no public condemnation of evolution (Artigas, 2006: 270–283).

STUDIED REJECTION OF DARWINISM: MARTIN S. BRENNAN

Of course, not all American Catholics were ardent champions of evolution. One reason is that American clergy in a dynamic and expanding frontier church had more pressing pastoral concerns than assessing arcane intellectual arguments they might not fully understand (Appleby, 1999:178). But even those who were attracted to the life of the mind, such as St. Louis seminary science professor Martin Brennan, did not always jump on the evolutionary bandwagon. Two years after the appearance of Zahm's book Brennan published *The Science of the Bible* (1898), a careful assessment of all the contemporary sciences in light of Scripture. Conversant with recent theory in physics, chemistry, biology, and geological "deep time," his sober review of the evidence led him ineluctably to the conclusion that Darwin's theory was simply wrong.

The weakest link in the chain of argument, in Brennan's view, was Darwin's confounding of "variety" with "species" (the same claim, curiously, that is made by some creationists today who concede micro- but not macroevolution). Brennan also regarded the incompleteness of the fossil record (which might hide missing links and connections) as damaging to the Darwinian hypothesis. Accepting the now-disproven argument of Sir William Thompson (Lord Kelvin) about the thirty-million year life of the sun, he concluded that solar physics offers good reason to believe that far too little time has elapsed to accomplish all the species diversification demanded of Darwin's descent by natural selection (Brennan, 1898:282, 286, 292). Brennan concluded that

[Darwin's] hypothesis, like all novelties and sensations in the scientific world, however popular and successful at first, is being tested in the crucible of facts and is declared a failure because it cannot satisfactorily answer the difficulties pressed against it. (Brennan, 1898: 294)

Thus, for Brennan, a special creative act was required at the origin of each species. Quoting Joseph Le Conte, Asa Gray, and Louis Agassiz, he denied the transmutation of species and contended that God had created the same number of species in the beginning as exist today. Moreover, regarding human evolution

Genesis tells us that God created man to his own image; ... Evolutionists rely upon biology and anthropology to establish their theories. But both biology and anthropology very plainly and positively favor the statement of Moses. (Brenann, 1898:315)

This proved to Brennan's satisfaction that "it is absolutely impossible for man to have been evolved by transmutation from any inferior species, but must have come by a special creative act of the almighty as the great Hebrew prophet records" (Brennan, 314).

After a further critical discussion of Darwin's and others' positions regarding evolution, Brennan concluded that "The truest results of biology and anthropology, instead of contradicting, confirm the Mosaic record. God called man into being by a special creative act. The whole human family belongs to the one same species, and man's Simian descent must be abandoned." He calculated human antiquity on earth at somewhere between 8,000 and 10,000 years. But that Brennan was not an unswerving Noachian literalist is demonstrated by his careful calculations and argument for a strictly local flood (Brennan, 1898:356; 385–386).

The difference in perspective between Martin Brennan and John Zahm is instructive. The fact that two scholars—both teaching in Catholic institutions in the American Midwest, and publishing within two years of each other—should come to opposite conclusions on the cogency of the theory of evolution, indicates the enigmatic role that personal predilection played in the halting Catholic acceptance of evolution.

A PROGRESSIVE VIEW FROM THE AMERICAN HIERARCHY: BISHOP JOHN L. SPALDING

In contrast to Martin Brennan stands John L. Spalding, Bishop of Peoria, Illinois, who in *Religion, Agnosticism, and Education* (1902) took as his starting point the intelligibility of the world: "We find that thoughts and things are coordinate. Ideas have their counterparts in facts. Everywhere there is law and order." Every aspect of animate nature unfolds by an inner drive, in which Spalding saw a divine planning:

In the minute cell there is the potency which creates the most perfect form. And, if it could be proven that the infinite variety of nature is but the result of the manifold evolution of a single elementary substance, we should still inevitably see the work of reason in it all. Hence when we know the world as an effect, we necessarily think of its Cause as having knowledge and wisdom; though the knowledge and wisdom of the Infinite are doubtless something inconceivably higher than what these terms can mean for us. (Spalding, 1902:95–96)

As a bishop, Spalding was in the delicate position of having both to represent faithfully the teaching magisterium of the Church and to reflect American values such as the separation of church and state and of academic freedom. In his fifth chapter, on "Education and the Future of Religion," he made a passionate plea for freedom of speculation:

To forbid men to think along whatever line, is to place oneself in opposition to the deepest and most invincible tendency of the civilized world. Were it possible to compel obedience from Catholics in matters of this kind, the result would be a hardening and sinking of our whole religious life. We should more and more drift away from the vital movements of the age, and find ourselves at last immured in a spiritual ghetto, where no man can breathe pure air, or be joyful or strong or free. (Spalding, 1902:175)

Spalding did not mince his words about the implications for education of a dogged adherence to outdated authority. He declared that "Aristotle is a great mind, but his learning is crude and his ideas of nature are frequently grotesque." Equally bluntly, he asserted that "Saint Thomas is a powerful intellect but his point of view in all that concerns natural knowledge has long since vanished from sight" (cited in Appleby, 1999:182).

Spalding noted that if Catholics hope to champion revealed truth effectively, they must be prepared to do so intentionally in light of modern scientific research:

If, in consequence, we find it necessary to abandon positions which are no longer defensible, to assume new attitudes in the face of new conditions, we must remember that though the Church is a divine institution, it is none the less subject to the law which makes human things mutable, that though truth must remain the same, it is capable of receiving fresh illustration, and that if it is to be life-giving, it must be wrought anew into the constitution of each individual and of each age. (Spalding, 1902:177)

But Spalding was equally aware of the dangers of reductionism, warning that overemphasis on scientific theories about the universe has created an atmosphere that attaches comparatively little importance to any factors other than heredity and environment:

The opinion tends to prevail that the mind and character of man, like his body, like the whole organic world, is the product of evolution, working through fatal laws, wherewith human purpose and free will—the possibility of which is denied—cannot interfere in any real way. (Spalding, 1902:198)

He asserted that no educator could accept this position without losing conviction in the ultimate and transcendent value of his work. But, "fortunately, one may admit the general prevalence of the law of evolution without ceasing to believe in God, in the soul, and in freedom" (Spalding, 1902:198).

GENETICS AND MENDEL THE MONK

There were many stages to the gradual assimilation of evolutionary theory into the framework of Catholic theology. In a real sense this parallels—if somewhat more slowly and with a different dynamic—the stages of its development within biological science. Darwin's idea of natural selection as articulated in 1859 was by itself an insufficient theory, for it could not explain why the variations upon which selection operated occur naturally within populations. For that, science had to wait for genetics, a piece of the puzzle supplied in part by another Catholic scientist.

Less famous than Darwin, although of no less significance for the success of the evolutionary theory, was Gregor Mendel (1822–1884), "father of modern genetics." Born in Brünn (then part of the Austrian Empire), Mendel studied at the Philosophical Institute at Olmütz, and entered the Augustinian Order at the Abbey of St. Thomas in Brünn (Martin Luther

Gregor Mendel (1822–1884). Augustinian friar whose research into the hybridization of pea plants helped to launch the modern science of genetics. (National Library of Medicine)

had belonged to the same order three centuries earlier). After studying at the University of Vienna, Mendel returned to his monastery to teach physics (Henig, 2000:40–46). As a scholar, he was inspired by his professors and his monastic colleagues to study the problem of genetic variation in plants, and over the course of nearly a decade, he cultivated and studied 29,000 pea plants. He discovered that in breeding hybrid peas, offspring in the first generation carried dominant and recessive traits in the ratio of 3:1. He presented his findings on "Experiments in Plant Hybridization" at the annual meeting of the Natural History Society of Brünn, and the paper was published in 1866 in the society's proceedings. Seriously criticized by the botanical establishment, the paper languished and was rarely cited over the next thirty-five years (Henig, 2000).

Mendel was elected abbot of his monastery in 1868, thus effectively ending his scientific career as he assumed an increasing burden of administrative and spiritual obligations. Providentially, his seminal paper on genetics was rediscovered in 1900 by the geneticists Hugo de Vries (Dutch) and Carl Correns (German), and thence it passed into history as one of the cornerstones of the neo-Darwinian synthesis. It was a Mendelian convert, the Dane Wilhelm Johannsen, who gave the name "gene" to the hypothetical unit of hereditary information. Mendel himself had referred to these units as "factors," characters and cells," and (considering his Catholic philosophical commitments) probably envisioned them as immaterial essences. It was only later that the laws of heredity were definitely shown to be carried by material chromosomes (Larson, 2004:162–166).

CONCLUSION: FROM THE GARDEN OF EDEN TO MENDEL'S GARDEN OF PEAS

For fifteen hundred years, the Christian community–including its clergy and theologians, its philosophers and scientists–had been committed to the only plausible account of life on earth, that narrated by Moses in the Book of Genesis. We have traced how this story began to lose its cogency, first in astronomy and physics, and then in geology and natural history, all the way to the discovery of genetics. We have also noted that there was considerable nuance to the various responses of Catholics in the nineteenth century to the theory of evolution. The pace of change in science and in other disciplines would only accelerate as the twentieth century dawned, provoking predictable responses. The numerous trends perceived by Rome as threatening to the established order were bundled into the term "modernism." These included Enlightenment secularism, a rationalist approach to biblical criticism, the replacement of Thomism with modern philosophical systems, and the historicizing view that the Catholic Church and its doctrines and practices can legitimately evolve over time. Pope Leo XIII had addressed the issue of critical biblical scholarship in *Providentissimus Deus* (1893), affirming the legitimacy of such an approach provided it was conducted in a spirit of faith. Pope Pius X launched a wider attack in his encyclical letter *Pascendi Dominici Gregis* (1907), lamenting that sacred studies (particularly in seminaries) were being neglected because the study of the natural sciences was so time consuming (Pius X, 1907:47). Meanwhile, judicious and balanced exegesis was being conducted quietly behind the scenes, as in Joseph Pohle's interpretation of the Genesis "days" of creation in light of expanding geological knowledge (Pohle, 1912: 117–123).

To early twentieth-century Catholics who were committed to Thomism—and the Thomist influence on seminary education can hardly be underestimated—Darwinian evolution seemed philosophically absurd, "deriving more perfect from the less perfect," being from nothing, an effect without a cause (Paul, 1979:62). H. Muckermann wrote in the *Catholic Encyclopedia* that an explanation of life "must ultimately be sought in a creative act of God, who endowed matter with a force *sui generis* that directed the material energies toward the formation and development of the first organisms" (Muckerman, 1907). And to demonstrate the fluidity of concepts in this discussion, William Hauber could on the one hand write approvingly of ornithologist Richard Goldschmidt's claim that it is consistent with the evidence to suppose that nature could produce "in one generation a full-fledged bird, ready for flight, and fit to survive in a hostile environment" (Hauber, 1942:162). On the other hand, he could note that "although St. Thomas had no reason to be an evolutionist, his theory of the tendency of matter to move toward greater and greater perfection in form was entirely compatible with biological evolution." Henry Woods firmly rejected the assumption that Augustine's doctrine of creation finds its logical conclusion in the theory of biological evolution (Woods, 1924:127–148), whereas Henry de Dorlodot agued that "the teaching of the fathers is very favorable to the theory of absolute natural evolution" (Dorlodot, 1922:169).

In the three quarters of a century of Roman Catholic deliberation about evolution before Teilhard's adoption of the theory, the focus of Catholic reaction was on two fronts: (1) the theoretical plausibility of the hypothesis and the extent of the supporting evidence—and here they mirrored to some extent the debate between scientists—and (2) the compatibility of the theory of descent with modification with the doctrine of creation in the Book of Genesis. Catholic thinkers spent relatively little time at this stage assessing the impact of evolution on doctrines such as theodicy, soteriology, Christology, or eschatology. This would be the ongoing task of theologians—Christian and otherwise—in the generations after the assimilation of the neo-Darwinian synthesis.

Zahm argued in his section on evolution and creationism for an admirable synthesis, of which no doubt Teilhard would be proud:

Then too, it will be manifest, that although truth was on the side championed by Aristotle, Sts. Athanasius, Gregory of Nyssa Augustine and Thomas Aquinas, by Buffon, Geoffroy Saint-Hilaire, Lamarck, Spencer, Darwin, Huxley, Mivart and their compeers, nevertheless, the opponents of the evolutionary idea, the Fathers and Schoolmen who favored the doctrine of special creation, the Linnaeuses, the Cuviers and the Agassizs, who resolutely and consistently combated evolution to the last, were all along but helping on and corroborating what they were intent on weakening and destroying. (Zahm, 1978:398)

Half a century later, Teilhard de Chardin would say in *The Phenomenon of Man*, that "religion and science are the two conjugated faces or phases of the same act of complete knowledge—the only one that can embrace the past and future of evolution so as to contemplate, measure, and fulfill them" (Teilhard, 283). Newman, Mivart, Zahm, and other early Catholic proponents of biological and human evolution would see as a logical extrapolation from their work Teilhard's claim that "[t]he human is not the static center of the world, as was thought for so long; but the axis and the arrow of evolution—which is much more beautiful" (Teilhard, 1975:7).

Chapter 4

Catholicism and Science at Mid-Twentieth Century

INTRODUCTION

It would be good to begin a chapter on the Catholic Church in the twentieth century by speaking about historiography, which is the view of how history should be written or understood. Often it is tempting to think about previous epochs in a negative light because of later events that seem so progressive by comparison. Such contrasts do exist of course, but they can often overshadow the frequent points of light that existed in what were otherwise dark periods. For example, the Second Vatican Council of Catholic bishops held between 1962 and 1965 overshadows the decades and centuries of Catholic thought and practice that preceded it. This gathering was a momentous three-year long series of meetings of the world's Catholic bishops in Rome. It is tempting to see the watershed changes of this council as an inevitable result of the Church's growing accommodation with modern life. Such a view sees this gathering of the Church's leaders as strictly a break from its conservative past. In such a reading, the Second Vatican Council is seen as a victory for a minority view that was, until that time, repressed by the church hierarchy. Consequently, when treating the Church's relations with science, it is tempting to think of some of the new language and thought at Vatican II as merely a result of the Catholic Church's new approach or even as an accommodation to science. While there is some truth to this characterization of events, it is ultimately too simplistic.

Is the modern Catholic Church favorable to science or wary of science from the twentieth century on forward? The answer to this question is difficult to summarize, but whatever one's interpretation of Vatican II, one can see how various Catholics as well as the Church itself have sometimes adapted to the spreading complexity of scientific advances and sometimes resisted them. True, there is much in the theology and intellectual climate of early twentieth-century Catholicism that is stifling and overly traditional. To a great extent then, the restlessness that gave rise to Vatican II can be seen as a reaction against this traditionalism in the Church.

Nonetheless, there are significant factors about Vatican II that can be better understood in light of developments that took place during the 1940s and 1950s. The relationship of the Church to the sciences in this period leading up to and including Vatican II is both one of hostility on the one hand, and harmony on the other hand. Vatican II came about for a variety of reasons, not the least of which were various fortuitous friendships forged among key bishops, the modern education of Catholic leaders, and especially the ebullient personality of Pope John XXIII. It is this Pope who called the Council shortly after he was elected in 1958. Science played a background role in shaping the awareness of many bishops, theologians, and church officials, who realized that the autonomy of the sciences from religious traditions was partly responsible for its success in making discoveries in the modern period. At the same time, there is evidence to suggest that the Church and its leaders realized the terrible, destructive and hence immoral capacities of science, as witnessed, for instance, in the nuclear bombings of Hiroshima and Nagasaki in 1945. The nature of the relationship between the Church and science in the mid-twentieth century can be summed up by saying that the Church rapidly increased its attention, understanding, and deliberations on the significance of the natural sciences and related technological advances.

CATHOLICISM AND THE NATURAL SCIENCES IN THE 1922–1965 PERIOD

Before we come to the ways in which Catholicism's relationship with the natural sciences evolved around Vatican II, it is important to investigate some of the intellectual trends, figures, and events that helped set the stage for Vatican II's new approach to the world, including the world of science. There are five main features of the 1922–1962 period that stand out. First, there is the struggle within Catholic theology over the kind of philosophy that could adequately account for both scientific study of nature while allowing the religious meaning of the world to be illuminated. At issue is whether or not Thomism as a philosophical system stemming from the thought of Thomas Aquinas (1220–1274) is vulnerable to the findings of the new physics and biology such as Einstein's relativity theory. As we will see, there were criticisms made from an early date against the

Pope Pius XII is the first Pope to acknowledge valid elements of evolutionary theory explicitly in a papal encyclical, *Humani Generis*. (AP Photo)

strictly logical, metaphysical emphases in Thomism since its revival in the nineteenth century.

Second, there are key doctrinal developments initiated by the Popes of this period, and in particular the encyclical *Humani Generis*, issued by Pope Pius XII. Third is the rise and growing significance of the Pontifical Academy of the Sciences. Fourth is the scientific work of Georges Lemaître whose research in cosmology has come to take on a special significance in light of Pope Pius XII's 1951 endorsement of the Big Bang hypothesis. Fifth, in light of the spectacular growth of knowledge in the biological sciences, we will examine briefly the thought and influence of Pierre Teilhard de Chardin (often referred to as simply "Teilhard"), a Jesuit and a paleontologist whose ideas came under critical scrutiny by the Church. The Church's initial criticism of his work was followed by a silence after Vatican II, at which time Teilhard's popularity increased dramatically.

The Catholic Church's relationship with the sciences in this period is mixed. On the one hand, one can see receptive views expressed about Darwinism, the most controversial scientific theory from a religious point of view in this period. In 1922, E.C. Messenger published an English translation of Canon Dordolot's 1909 book *Darwinism and Catholic Thought*, in which Dordolot, an eminent cleric from the prestigious Louvain University in Belgium, praised Darwin's theory from a Catholic viewpoint. On the other hand, we see during the 1940s and 1950s that the Catholic press and popular opinion amongst Catholics turned more hostile toward evolutionary theory. In fact, it has been observed that Catholic opinion in the 1950s toward evolution was more hostile than the Catholic reaction in the decades immediately after Darwin's theory was first propounded. What can be cited as reason for this tendency?

THOMISM BEFORE VATICAN II

The role of the officially mandated school of philosophical thought, what is called "neo-thomism" or simply "Thomism" is very complex to assess in terms of the Church's relationship with science. Thomism possesses both positive and negative attitudes toward science that are difficult to sort out because of the growing pluralism in philosophical interests taken by Thomists since the 1960s. Catholics marvel that a philosophical system that seemed to be so stable before the 1960s almost ceased to be taught in many Catholic seminaries after Vatican II. More than just the Council's interest in pastoral concerns and its historical appreciation of human cultures are at work in breaking Thomism's monopoly.

Thomism's demise can be interpreted in several ways. One reason attributed is its alleged lack of rhetorical skill. In 1942, Robert E. Brennan notes a frequent criticism of Thomists is that "matter and content they have in abundance, but good literary form is almost an unknown quantity" (Brennan, 1942, 21–22). The problem is that Thomism had become caught in a logical system that was inherently abstract. Thomists hold to a system of distinctions among both the bodies of knowledge as well as the natural objects that these forms of knowledge are concerned with. The poor literary quality of Thomism can be traced to its many-layered interpretations of Aristotle's metaphysics.

Aristotle himself developed a relatively simple worldview. He distinguished the natural world in two ways. First, Aristotle distinguishes between a thing's essence and its accidents. Second, the relationship between a thing's essence and its accidents is an account of how it changes. Change can be analyzed and broken down according to three constant elements: a thing's potency, its form, and its act. Brennan contrasts Thomist metaphysics with the philosophy of early twentieth-century thinkers like

George Santayana, William James, and Aldous Huxley. As opposed to the Thomists, these philosophers gracefully yet forcefully opposed the idea that metaphysical essences lie "behind" material things. Santayana, for example, sees material reality as primary. Any talk of God or form is merely intended to promote the *illusion* of a purposeful universe. Science, with its accent on mechanical and machine-like structures, came to be assumed as the basis for such alternatives to Thomism.

William James, who is widely known as the "father" of American pragmatist philosophy, was himself very interested in religion. But his interest in religion and God differs fundamentally from Catholic philosophy. James was interested from the start in how religion is practical and experiential. He was not interested in doctrine or church teaching. His famous work *The Varieties of Religious Experience* deals not with abstract philosophical ways to understand God's action in the world, but instead with how God is experienced as an agent of change in one's personal life. So, we can see why, in contrast with other philosophers such as most Thomists, even one sympathetic to the basic aims of Thomism during this period would be skeptical about that philosophy's effectiveness.

But, there were more serious problems with the kind of concepts and language Thomists used to articulate a view of God and the world of science. The physical causes of empirical phenomena are the paramount concern for twentieth-century scientists and empirical philosophers. The attempt to categorize causes metaphysically they regarded as nonsense or redundant verbiage. From a Catholic perspective, without metaphysics, science and other forms of knowledge would be fractured. As a consequence, Catholic philosophy of science tended to see a prominent role for philosophy in explaining scientific epistemology, as opposed to certain scientific concepts of nature. Catholic priest and philosopher of science William Wallace, O.P. summarizes it from a philosophical point of view:

[T]wentieth century Catholics have shown little interest in science, being concerned mainly with metaphysics and social and political thought. Notable exceptions are Jacques Maritain, Charles de Koninck, and Vincent Edward Smith, all of whom developed philosophies of science. All three . . . were too much influenced by Duhem and the Critique of Science Movement, and tended to deny to modern science the possibility of attaining demonstrations in the world of nature. A moderate realist position that allows such a possibility seems more in accord with Aquinas's own thought. (Wallace, 2001:452)

Once again, we come across the name of Thomas Aquinas. According to Wallace, Thomism was already in decline for the sheer reason that his philosophy was understood to be less relevant to Catholic philosophers and scientists. This contrasts with the official position of the Catholic Church that continued to view Thomas' philosophy as pre-eminently able to mediate or interpret science in a way conducive to Catholic thought.

Duhem's avoidance of Thomism, which was discussed in chapter two, is an instance of the flight from Thomism. The main reasons for the lack of confidence in Thomism during this period are twofold. First, Thomists were reluctant to emphasize probability. They preferred to think of knowledge in terms of certainty. Second, Thomists were late in endorsing indeterminism and randomness at the quantum or subatomic level of the universe. Catholic philosophers would rectify these inadequacies from different Thomist perspectives later on. But, in the period between the First World War and the Second Vatican Council, it did not seem as if Thomism was intellectually satisfying for those philosophers who were familiar with the advances of the new sciences, especially physics.

Relativity theory and quantum theory are the most important twentieth-century scientific challengers to Thomist philosophy. Following the Thomist revival of Leo XIII, Thomists reiterated their belief that sense knowledge could yield certitude. That is, certain knowledge of things was thought to be fixed, in part because of the absoluteness of space and time. Special relativity theory challenges this assumption because after Einstein's discovery, time becomes relative to the speed of motion. Therefore, it is no longer possible for there to be a single, universally accessible observation of an event. In physics, observations are now dependent on the location and speed of the observer as well as the object being observed. Of course, Thomism is only one of a number of philosophical schools of thought affected by the changes wrought by relativity theory.

However, quantum mechanics would throw an even more dramatic challenge to Thomism because of its theory of nonlocal action and the elusive behavior of subatomic particles. Here, the challenge to Thomism is threefold. First, it is ontological. The Thomist belief in essences or "thingness" flounders. For example, is an object being studied a wave or a particle? If the thing or entity being studied does not possess a stable identity, Thomism appears inept to account for this entity metaphysically. Second, it is aetiological. Aetiology is the study of causes, and in quantum events or structures, causality has to be totally reframed. In classical science, Thomism understood the underlying cause of a phenomenon to be known when the cause states or implies a precise time for the phenomenon's occurrence. In quantum structures, this precision is denied. In its place come predictions of the range of all possibilities, not just one specific instance. Only statistical forms of knowledge are possible at the quantum level, and not until the 1950s did a few Catholic thinkers really grasp this fact. Third, quantum physics also challenges Thomism at the epistemological level. According to the Heisenberg Uncertainty Principle, the position and the momentum of a particle cannot be simultaneously identified. The effort to determine one of these features of a particle carries with it the inherent inability to determine the other feature. At the subatomic level, it is impossible to measure subatomic particles in their total identity. Certitude is not just very difficult because of the sheer number of factors to be considered, as in chaotic systems like weather. Certain knowledge about the subatomic level of reality becomes inherently impossible, because the objects of that level simply cannot be measured.

These three aspects of quantum physics and relativity theory have been especially difficult but not impossible hurdles for a Thomist philosophy of nature to overcome. Against idealists and social constructionists who think knowledge is what it is that human subjects authorize through their concepts, Thomists argue for objectivity in knowledge. From Aristotle's four kinds of causality (efficient, material, formal, and final) Thomists have always valued sensation and the knowledge of causes as responsible for change in these ways.

At Vatican II however, an ambivalence surrounding Thomist thought set in at seminaries and other Catholic institutions of high education. Vatican II's *Decree on Priestly Formation* states: "Basing themselves on a philosophical heritage which is perennially valid, students also should be conversant with contemporary philosophical investigations, especially those exercising special influence in their own country, and with recent scientific progress" (Abbott, 1966:452). Pluralism in philosophy was introduced much more explicitly than it had ever been adopted. Science, in the way Vatican II envisioned it, was autonomous from Thomist philosophy. The result, when combined with the drop in the number of seminary students in western countries after the 1970s, is a decline in philosophical formation in favor of pastoral formation. Because there has been a dramatic rise in the proportion of seminary students from non-Western cultures in Africa, Asia, and Latin America, Thomism has lost a tremendous amount of prestige. While its decline has since been arrested and Thomism has undergone something akin to a revival recently, the Council's embrace of dialogue with other philosophical positions secured Thomism's loss of primacy as a systematic philosophy through which the Church and Catholics could unflinchingly absorb the advances of science.

PAPACY AND DOCTRINE

The papacy of Pius XII (1939–1958) is of the utmost importance in understanding the ambivalence of twentieth-century Catholic attitudes toward science. In many respects, his nearly twenty-year leadership of the Catholic Church represents the last vestige of strict monarchical approach to church government. In 1937, the papal encyclical *Divini Redemptoris* had been authorized by the previous pope, Pius XI. This document prohibited Catholics from giving any help to the Bolsheviks. This was typical of the sort of didactic anti-Marxist outlook prevalent at the time, tempered as it was by the authentic desire to protect Catholic and Christian values and culture that were perceived to be under attack.

Yet, it is under Pius XII that we see the first openings to a genuine autonomous autonomy for the world of science. One reason underlying this opening is likely the rising participation by Catholics in science. Indeed, this fact was cited in apologetical works of the period. One such work is Bertram Windle's 1927 book *The Catholic Church and Its Relations with Science*. Partly in response to the anti-Catholic sentiments expressed by Thomas Huxley, a prominent British atheist and partisan of Darwinian natural selection, Windle comments:

> ... what of Galvani, Volta, Ampère, Coulomb, Ohm, all of whose names are now immortalized in the nomenclature of electricity? What about the Abbé Haüy, the father of crystallography, who died the same year as Pasteur, who began his triumphant scientific progress by investigations along lines made possible by the researches of the Abbé? Again, what shall be said of Nicolaus Stensen, the father of modern geology, a convert from Lutheranism who died a Catholic bishop? ... when it comes to matching man of science with man of science the Catholic side will not be the first to run out of names. (Windle, 1927:45)

Here is an argument that was not untypical of the period: Catholic participation in science implied Catholic acceptance of science, provided that science was not what atheists like Thomas Huxley thought science implied for religion. For Windle and the Catholic Church, Catholic acceptance of science should have meant the scientific community's acceptance of the Church. But for Huxley and John William Draper, author of the 1874 book *History of the Conflict Between Religion and Science*, Catholicism seemed an institution steeped in superstition and antiscience prejudice.

Due to both the greater participation of Catholics in science and the influence of scientists in the Pontifical Academy, the Catholic Church began to make substantial changes in the way it viewed knowledge in scientific disciplines. But, originally this opening did not come through any consideration of biological or cosmological theory. For reasons that pertain to the Church's search to clarify God's revelation in history, the Church made its first openings to scientific research in the area of biblical research.

After a period of antimodernist reaction against liberalism within the Church in the early part of the century, new institutions were born that would later play a key role in opening up the Church to more liberal views. In 1909, Pope Pius X established the Biblical Institute in Rome under the watchful eye of the Pontifical Biblical Commission. The Institute, staffed by Jesuits, was mandated with the teaching of biblical exegesis, the theological task of analyzing texts to understand the parameters of valid interpretations. The Biblical Commission, which was established to serve as the Church's official organ responsible for communicating conclusions on matters of scriptural interpretation, responded to an attack made against the teaching of scripture at the Biblicum by one Dolindo Ruotolo. Ruotolo had attacked the exegesis being done at the Biblicum, but in responding to this attack, the Biblical Commission succeeded in setting out more liberal criteria for interpreting scripture than was previously imaginable. Parts of the text of the Commission made their way into the 1943 encyclical *Divino Afflante Spiritu* (Walsh, 1991:22).

This is a significant event in the history of church–science relations, for even though the subject matter is not scientific per se, it did indicate the Church's acknowledgement that scientific investigation has a significant impact on the way in which church teaching can be communicated. In this document, the Church grants permission for biblical scholars to study biblical texts in their original Hebrew and Greek languages, not Latin, which is a later translation of both the Old and New Testaments. Catholic biblical scholars also received permission to use the now popular historical-critical method of textual study. The significance of this encyclical is hard to underestimate. It is an indicator of the church's developing approach to scientific methods and empirical knowledge in general. In effect, more leaders in the Church saw that the autonomy and content of the historical sciences, even those bearing on scripture, did not necessarily threaten the church and its teachings. The implications of these developments were that Catholic scientists would enjoy greater freedom in the future with respect to any research that related to church teaching.

Indeed, in 1955, the secretary of the Pontifical Biblical Commission stated that biblical scholars would enjoy freedom in their research to the point of not being bound by the strictures laid down during the 1905–1915 period, such as the Church's requirement that Catholic scholars hold to the Mosaic authorship of the entire Pentateuch (that Moses wrote the first five books of the Old Testament or Hebrew Bible himself). Neither would Catholic scholars be bound to believe that the first few chapters of Genesis are a correct historical chronology of earth history. In 1964, the Commission issued an instruction on "The Historical Truth of the Gospels" that carried over insights from historical research into the four New Testament gospels. The Commission acknowledged that the gospels were written in stages, a claim that they made strenuous efforts to suggest does not affect negatively the truth of the gospels (Rausch, 2000:58–59).

Humani Generis

However, the hopes of those scientists and philosophers who wished for a church that treated science with more perspective were somewhat dampened by Pius XII's encyclical *Humani Generis*. Released on August 12, 1950, it is subtitled "an encyclical of Pope Pius XII concerning some false opinions threatening to undermine the foundations of Catholic Doctrine." It contrasts the relativism of various philosophies, including certain ways of interpreting evolution, with the Church's teaching authority. As such, it is apologetic and it possesses a key significance for church–science relations during the period.

Pius XII reinforces the claims of Thomism through the legitimacy of a church teaching authority, which by its nature must mediate the truth of scripture for the purpose of interpretation. In the following passage, Pius XII speaks of the errors of Biblicism, an approach to revelation that takes scripture alone to be authoritative, and in so doing, undercuts the authority of reason. With reference to an ambiguous number of former rationalists, he states that there are

> former adherents of rationalism today [who] frequently desir[e] to return to the fountain of divinely communicated truth... But... not a few of these, the more firmly they accept the word of God, so much the more do they diminish the value of human reason, and the more they exalt the authority of God the Revealer, the more severely do they spurn the teaching office of the Church, which has been instituted by Christ, Our Lord, to preserve and interpret divine revelation. This attitude is not only plainly at variance with Holy Scripture, but is shown to be false by experience also. (*Humani Generis*, #8)

At the same time, he pillories the philosophical systems of "historicism," "existentialism," and "evolutionism." These philosophical systems had attracted growing numbers of adherents in European universities during the first two thirds of the twentieth century.

No theologians or philosophers were identified by name in the encyclical. Yet, "evolutionism" is thought to refer specifically to the French Jesuit and palaeontologist, Teilhard de Chardin, whose integration of scientific research with theological elements was unique and, many would say, visionary. Teilhard represents, for Pius XII at least, the possible threat of a new modernism, an internal threat to the intellectual cohesion and unity that neo-Thomism had provided the Church, at least since *Aeterni Patris*. Other theologians are also thought to have been the targets of this encyclical's strictures, including Marie-Dominique Chenu, Henri de Lubac and Karl Rahner.

In *Humani Generis*, Pius XII denounces those who interpret biological evolution as a total theory of human origins. Officially, for the Church at this time, evolution is still perceived as a theory that potentially denies an arena for divine action. Being written in the mode of defensive apologetics, *Humani Generis* takes a more overtly hostile stance against evolution than previous church documents. It questions evolution largely from the perspective of epistemology, and bears an undeniable likeness to the traditional views expressed in defense of church authority. To the extent that the Church is simply defending its purview, its views on evolution are not extraordinary. In the encyclical, for example, it is written:

In theology some want to reduce to a minimum the meaning of dogmas; and to free dogma itself from terminology long established in the Church and from philosophical concepts held by Catholic teachers, to bring about a return in the explanation of Catholic doctrine to the way of speaking used in Holy Scripture and by the Fathers of the Church. They cherish the hope that when dogma is stripped of the elements which they hold to be extrinsic to divine revelation, it will compare advantageously with the dogmatic opinions of those who are separated from the unity of the Church. (*Humani Generis*, #14)

While the subject of this particular section of the encyclical is the nature of concepts that communicate the meaning of divine revelation, the impression that is given, is that Christian scientists such as Teilhard are in error when describing evolution as the mode of God's creation of the universe. The encyclical states famously:

... caution must be used when there is [the] question of hypotheses, having some sort of scientific foundation, in which the doctrine contained in Sacred Scripture or in Tradition is involved. If such conjectural opinions are directly or indirectly opposed to the doctrine revealed by God, then the demand that they be recognized can in no way be admitted. (*Humani Generis*, #35)

The overtones of the Galileo case are unmistakable. Yet, in the next section, the text allows that the theory of evolution may be researched and discussed "in as far as it inquires into the origin of the human body as coming from pre-existent and living matter." Evolution however is precisely the kind of hypothesis that should be treated with caution, for while the idea that the human body originates from other nonhuman living beings is not problematic, nevertheless the encyclical states that any hypotheses which oppose the direct creation of the human soul by God are to be rejected. Only God is involved in the creation of the soul, and any scientific theory that proposes to do away with the soul, or explain it in naturalistic terms, is contrary to church doctrine according to Pius XII.

This is also the case with polygenism, the belief that the first human beings number more than the two first parents mentioned explicitly in the

Bible: Adam and Eve. Again, the encyclical is clear regarding this liberal way of treating the biblical text:

> ... the faithful cannot embrace that opinion which maintains that either after Adam there existed on this earth true men who did not take their origin through natural generation from him as from the first parent of all, or that Adam represents a certain number of first parents. (*Humani Generis*, #37)

Almost as a holdover from the biblical literalism of the previous century, this encyclical holds that Adam and Eve cannot be conceived in a symbolic manner. They are real individuals from whom all others are descended. This is the view known as monogenism. The reason for maintaining this stance was the perceived threat to the doctrine of original sin: if Adam and Eve were taken to be *symbolic* figures rather than humanity's actual common ancestors, it becomes more difficult to affirm that original sin is transmitted through the generations, making it a universal feature of the human condition. The doctrine of original sin has been conceived as an inherited feature of human existence transmitted through sexual intercourse and the conception of a human person. Saint Augustine is the most influential figure in elaborating this doctrine during the fifth century. Based on an interpretation of the important biblical passage, Romans 5:12 ("Sin came into the world, and death through sin, and so death spread to all men, through one man, in whom all men sinned"), Augustine saw the individual man, Adam, as a necessary element in accounting for all of humanity's sins. Through "carnal begetting" as Augustine put it, all humans have a connection to the sin of Adam through its transmission from generation to generation. Potentially, just as sin came into the body through a decision of Adam's human soul, so redemption is won through Jesus by our decision to acknowledge Jesus through faith. Faith has, as its eventual reward, the body's resurrection at the end of time. Such is the standard Christian narrative. However, the thrust of this theological outlook has been maintained in contemporary Catholic teaching, for instance in the Catechism, but it now includes references to the Bible's "figurative" language, and no reference to the need to hold for a belief in one and only one original person named Adam.

Staying with a cautious interpretation of evolutionary theory, in section five of *Humani Generis*, the world's evolution is regarded as impermanent and not primary:

> Some imprudently and indiscreetly hold that evolution, which has not been fully proved even in the domain of natural sciences, explains the origin of all things, and audaciously support the monistic and pantheistic opinion that the world is in continual evolution. Communists gladly subscribe to this opinion ...

The political thrust of the Church's concern is clear. And, like communism, existentialism is also tied to an excessive enthusiasm with evolutionary theory. According to the document, existentialism is faulty since "it concerns itself only with existence of individual things and neglects all consideration of their immutable essences" (*Humani Generis*, #6). So, this encyclical maintains an opposition between Thomism and existentialism. In summary, evolution is a key to modernity's ills, since it gives change priority over order and stability, whether in nature, society, or philosophy.

THE PONTIFICAL ACADEMY OF SCIENCES

With the issuing of a *motu proprio* (or administrative papal bull) on October 28, 1936, Pope Pius XI reestablished an ancient academy, the "Lincei," as the new Pontifical Academy of the Sciences, the *Pontificia Academia Scientiarum*. "Dei Nuovi Lincei," as its predecessor was called in Latin, served as a council of sage advisors. The present academy has come to serve as the Catholic Church's most high-profile body of scientific researchers upon whom the Pope and bishops rely for authoritative knowledge about scientific issues. More recently, as the Pope stated in a recent address, the Academy has focused on humanistic and moral issues as these arise in cultural contexts. The Academy's first president was an Italian, Agostino Gemelli.

Many are surprised to learn that not all academicians in the Pontifical Academy are Catholic. In fact, in order to reserve for itself the most updated and expert views on scientific affairs, the Academy has diversified from its beginnings to include non-Catholics. Among the original members of the Academy who were named on October 28, 1936 are Max Planck (a German physicist whose work in quantum theory generated a wholly new approach to atoms and subatomic particles) and Erwin Schroedinger (an Austrian physicist whose model of the atom reflected insights developed by Einstein and Planck). Both Catholic and non-Catholic members of the Academy are still among the most well known members of the world's scientific élite.

Apart from the fact that Academy meetings have become more regular since World War II, there has been at least one discernible change in Academy activity. Beginning in the 1960s around the time of Vatican II, the Pontifical Academy has placed more effort in probing the moral and ethical implications of modern technology in its study weeks and various congresses. Various meetings or "Study Weeks" have been held, for example, on the subjects of "The Use of Fertilizers and their Effect on the Increase of Crops" (1972), "Biological and Artificial Membranes and Desalination of Water" (1975), "Natural Products and Plant Protection" (1976), "Energy and Humanity" (1980), and "Modern Biological Experimentation" (1982).

Academy activity in moral issues has reflected the growing moral content of church teaching generally, especially under the papacy of John Paul II.

GEORGES LEMAÎTRE

Georges Lemaître was born in Carleroi, Belguim in 1894, and began his scientific career at the College of Engineering in Louvain. After time spent in the Belgian artillery in World War I and seminary training where he continued to read mathematics and science, he travelled to Cambridge University in England. At Cambridge, Lemaître studied Einstein's General Theory of Relativity that had been discovered ten years earlier but which had not been adequately interpreted. At this point in the 1920s, it was still unclear what kind of universe the General Theory of Relativity implied. All that was known for certain was that the theory predicted a relationship between space and time, and in addition to that, a relationship between space-time (as we call it now) and the amount and distribution of an object's mass. Einstein had formulated his theory around 1915, but it was unclear how its predictions resulted in the kind of universe we observe. In contact with the well known Cambridge astronomer Arthur Eddington, Lemaître proposed in 1927 that the universe was expanding. Einstein did not think much of this theory at first and told Lemaître "your calculations are correct, but your grasp of physics is abominable" (Midbon, 2000:18–19). In 1931, after Eddington had adopted Lemaître's expanding universe theory, Lemaître revised it to state clearly that the universe evolved from a definite beginning, not just a preuniverse of mass as predicted by Einstein's theory. At this point, Eddington rejected the concept of the universe's absolute beginning as "repugnant," but Einstein embraced it in 1933 after Lemaître clarified that the world could have begun as a single quantum, a state in which the notions of space and time break down. It was on quantum mechanics that Lemaître seemed to pin his hopes for a later, fuller explanation of the primordial or "primeval" atom, the original single quantum. On the basis of his pioneering scientific work, Pope Pius XI inducted Lemaître into the Pontifical Academy of the Sciences in 1936.

This hypothesis of a primordial beginning point became known and is still known as "the Big Bang," and was popularized as such by George Gamow, another astronomer. The phrase "Big Bang" also came to be used as a term of ridicule by Cambridge astronomer Fred Hoyle, an atheist. But, ironically, the choice of wording stuck. Both advocates and critics have been using the term ever since. In Big Bang theory, to summarize, the universe is understood to be a single reality of space and time that is constantly expanding. The rate of its expansion and the cause or causes of its expansion would later become the primary points of contention once confirmation of its expansion was made in the 1960s. In 1965, the expansion of the universe was discovered in a new way through the measurement of the cosmic background radiation that was left behind by the Big Bang. This measurement by Arno Penzias and Robert Wilson calculated a uniform temperature of the cosmic radiation at 2.7 degrees Kelvin. Since this measurement, which earned Penzias a Nobel Prize in physics, the Big Bang has came to be widely accepted, almost a consensus among cosmologists—this after it languished in obscurity throughout the 1920s and 1930s. Lemaître heard of Penzias and Wilson's discovery shortly before he died, and this must have been heartening news given earlier, unresolved tensions with Einstein over the rate of the universe's expansion and the ongoing hostility toward Lemaître by Cambridge astronomers like Fred Hoyle and other adherents to a competing, "Steady-state" model of the universe.

In connection with the Big Bang hypothesis, Pope Pius XII is responsible for possibly one of the most controversial episodes in the twentieth-century relationship between Catholicism and the natural sciences. On November 22, 1951, he delivered an address to the Pontifical Academy of Sciences during one of the Academy's study weeks. In this speech, Pius XII takes the relatively unusual step of advancing a theological interpretation of the Big Bang hypothesis of cosmic origins.

By the early 1950s, the Big Bang was a scientific hypothesis with a growing number of adherents, but without the empirical verification that allowed its defenders to boldly confirm that it was universally acknowledged to be verified. Nonetheless, Pius XII concludes his address to the Academy in the following way:

What, then, is the importance of modern science in the argument for the existence of God drawn from the mutability of the cosmos? [Science] has widened and deepened to a considerable extent the empirical foundation upon which the argument is based and from which we conclude a self-existent Being immutable by nature ... it has followed the course and direction of cosmic developments ... it has indicated their beginning in time at a period about five billion years ago, confirming with the concreteness of physical proofs the contingency of the universe and the well-founded deduction that about that time the cosmos issued from the hand of the Creator. (Martini-Bettòlo, 1986:83)

What is immediately noticeable from the standpoint of later cosmology is the fact that the Big Bang hypothesis is now couched in terms of a singularity at the origins of time and space at roughly *thirteen* billion years ago, not the *five* billion referred to by Pius XII. The other feature about this discourse is the word "proof," a word that has long since disappeared in most of scientific writing, for its association with solid certainty. While it is still used by many scientists, the word "proof" is highly contentious.

The difference between the five billion years of early Big Bang theory and the later hypothesis of thirteen billion years shows just how risky it is to base a theological conclusion on specific data and even specific theories in cosmology and other sciences as well. In fact, it is this very modification of the Big Bang hypothesis that caused an uproar among a number of scientists, especially Gamow and Lemaître. Lemaître is reported to have been publicly agitated with the Pope over the latter's endorsement of the Big Bang in such a simplistic way. He believed that the Pope's particular endorsement of the Big Bang as the creation caused two other serious problems, one theoretical and one political.

In response to the Pope, Lemaître speculated that there could have been a period of cosmic contraction that preceded the Big Bang and the ensuing expansion. This theoretical possibility, while difficult if not impossible to verify from the standpoint of us human observers thirteen billion years on the "other side" of the Big Bang, would cause tremendous difficulties for the kind of theological statement about creation that Pius XII sought to reinforce.

The more political point has to do with the emerging debates in the 1950s between Lemaître and other advocates of the Big Bang hypothesis on the one hand, and atheist cosmologists and physicists on the other hand. The latter scientists' work was largely based on a Marxist atheism that is hostile to the Big Bang for its perceived benefits to Christian belief. For Lemaître, it was essential to maintain a purely scientific basis for the Big Bang hypothesis as it would inevitably meet with criticism and verification. For it to be adopted so readily by the Pope would be, for his Soviet and atheist counterparts, sufficient evidence to identify the theory's religious roots and destroy its scientific credibility (McMullin, 1981:54). Thus, during the papacy of Pius XII, there were genuine fissures in Catholic appraisals of science, which were serious disagreements in philosophical and theological approaches. They did not become open conflicts, yet later developments under future Popes more effectively prevented these disagreements from flaring into major conflicts. What this episode underlines is the singular importance of Lemaître as a leading Catholic scientist of the twentieth century whose efforts should be seen as influential on church understanding of science. Specifically, Lemaître's efforts probably made sure that the Church would be more careful in the future in its public reflections on scientific questions.

TEILHARD DE CHARDIN

Teilhard de Chardin is perhaps the most famous Catholic scientist of the twentieth century. Born in 1881 and raised in central France, Teilhard grew up in a traditional Catholic milieu, and was encouraged by his parents from

Père Pierre Teilhard de Chardin, consulting Paleontologist of the National Geological Survey of China, is shown at a symposium of Early Man at the Academy of Natural Science in Philadelphia, Pa., March 18, 1937. De Chardin holds a skull of a Peking man he found in Peking, China. (AP Photo)

an early age to pursue a religious devotion to Our Lady and the Sacred Heart of Jesus. Imbued with a mystical appreciation of nature, Teilhard collected specimens of rocks, fossils, and other specimens of nature as well. He joined the Jesuits at the age of eighteen, with whom he felt that he could pursue both his religious and scientific interests. However, the Jesuits were expelled from France shortly thereafter. After a period of theological study in England, he returned to scientific studies in Paris until World War I broke out, whereupon he served as a stretcher bearer in the trenches. After the war, further study eventually led to a position in Geology at the *Institut Catholique de Paris*, where he faced his first difficulties with the Church over his increasingly public stance regarding the interpretation and meaning of evolution.

Inspired at first by Henri Bergson's philosophy, and after reading Bergson's book *Creative Evolution*, Teilhard became convinced that evolution was a basis not only for understanding the foundations of biology, but also for understanding God's relations with the world as a whole. In this vein, by the time he wrote his widely read books, *The Divine Milieu* (1927) and *The Phenomenon of Man* (1938–1940), Teilhard had developed a sophisticated metaphysical theology of creation. In contrast to the idea of the hyper-competitive "survival of the fittest" as we know it from some interpretations of Darwinian theory (such as social Darwinism), Teilhard interpreted evolution as a process heading toward ever greater complexity and toward higher forms of consciousness and eventually toward an "omega point." This may be said to be Teilhard's leading idea.

Consisting of a number of discrete steps in the creative process, Teilhard recognized what he saw as the divine, creative link between *cosmogenesis*, *anthropogenesis*, and *noogenesis*. These three "geneses" refer to the emergence, respectively, of the cosmos, of humanity (the anthropos), and the human mind or the *nous*, which is the ancient Greek word for "mind." Human intelligence, Teilhard notes, emerges as a mind-sphere or *noosphere*. Recently, some have come to see cyberspace, the Internet and related technologies, as the further extension of the mind-sphere. So in a sense, Teilhard's prescience regarding evolution and its extensive meaning may be more significant than previously thought.

The creative links between the emergence of cosmos, human life, and intelligence met a limit however, which for Teilhard is the limit of redemption. Nature and culture are separated realities. Since humanity is unable on its own to be redeemed, there is the further (supernatural, as opposed to natural) gift of *christogenesis* that only comes from the recognition of Christ seen with the eyes of faith. Only with the eyes of faith are we able to see what God has to offer above and beyond the extraordinary diversity of natural creation and human intelligence.

As we saw above in reference to the papacy of Pius XII and the encyclical *Humani Generis*, the Catholic Church initially opposed Teilhard's views on evolution, and his work was caught up in the general conflict between historical, scientific study and church authorities during the post-World War II period. Beginning in the 1950s, Teilhard's scientific and religious writings were embraced by many Catholics, especially those whose views were more liberal or were predisposed toward the modern world. For these people, Teilhard represented an open outlook toward the world, an intellectual link between Catholic faith and the secular world. This outlook promised that one did not have to give up one's allegiance to the Church for that of the world and vice versa. But, Teilhard suffered a great deal of personal anguish over the fact that he was not allowed by his Jesuit superiors to publish his works earlier, as they were originally written in the 1920s and 1930s. One of the main reasons that he was forbidden to teach paleontology, his research specialty, in France after World War I, was because Teilhard understood the human species as linked to the global process of biological evolution, and this was frowned upon.

On June 30, 1962, the Holy Office of the Catholic Church issued a *monitum*, which warned Catholics against accepting Teilhard's theories uncritically (AAS, 1962:526). While this warning was interpreted by many as a condemnation of Teilhard's thought, in fact it was neither a condemnation nor even a listing of his works in the famous *Index of Forbidden Works*. Nevertheless, the warning had a profound impact on the way that his thought would be henceforth received. It was around this time that Rome temporarily halted the publication and translation of theologian Henri de Lubac's study of Teilhard's thought. This warning confirmed for many that the *possibility* of a modern Catholic form of antievolutionism would continue to be defended by the Church, in spite of its imminent turn toward the modern world.

From an early analysis of Teilhard by de Lubac (who later became a cardinal in the church), we know that Teilhard was interested in science for apologetical purposes:

> ... the entire work of Père Teilhard de Chardin can be regarded as one vast proof—renewed in a scientific perspective—for the immortality of the human soul and the existence of God; a proof that is completed by a preparatory effort reaching to the threshold of the Christian faith. (de Lubac, 1968:11)

Without a doubt, Teilhard later became the leading figure for those Catholics who sought to reconcile faith and science. Today, conferences, journals, and academic institutes are named in his honour and scholarly events are organized concerning his thought. In fact, de Lubac himself, a Jesuit priest who became increasingly conservative in his later years, was asked by his Jesuit superior to examine Teilhard's thought. De Lubac's theological and philosophical acumen was peaked by Teilhard's obviously extensive knowledge and compelling spiritual vision. At first, for those in authority in Rome, Teilhard's name was associated with liberal excess and theological error. This is why his early writings were condemned, associated with the Index of a list of erroneous or scandalous works that was maintained until it was dismantled around Vatican II.

What made Teilhard's life even more difficult was the revelation, in 1953, that the 1912 archaeological discovery of "Piltdown Man" was in fact an elaborate hoax set up by an amateur collector of anthropological artefacts, one Charles Dawson. This story can be summarized as follows: Teilhard had met and corresponded with Dawson, an amateur palaeontologist, at around that time in England where Teilhard was studying. The 1912 discovery was supposedly that of a skull and jaw bone belonging to a previously unknown species of *hominid* man. In fact, as the 1953 investigation uncovered, the skull was of a recently deceased human being, and the jaw bone belonged to an orangutan. Piltdown man was a hoax.

Teilhard's alleged association with Dawson in this unseemly affair was made worse by magazine columns written by none other than the famous science writer, Stephen Jay Gould. Gould tried to discredit Teilhard for plotting the hoax in cooperation with Dawson.

What Gould seems to have sought, in making these criticisms, was to show that Teilhard's supposed theological interest in multiple lineages for the emergence of human life dictated Teilhard's science. As one famously agnostic about God and about scientific evidence for God's existence, Gould thought he saw in this incident a case where religious belief in spirit over matter was not scientifically credible. Gould also uses the word "cult" in referring to Teilhard's followers. But, Teilhard's innocence has since been upheld by historians and others who have uncovered no motive for Teilhard engaging in a hoax. Nor is there any evidence from the chronology of events from 1910 to1914 that could support Teilhard's involvement in Dawson's joke (King & Salmon, 1983:159–169). This has not stopped other agnostic scientists, namely Daniel Dennett, from engaging in similar attacks. On the other hand, Teilhard has been criticized by religious traditionalists for neglecting the supernatural in his embrace of Christ:

Herein lies the danger of Teilhardism. Its emphasis on the incarnational and cosmic Christ, to the detriment of the redeeming Christ, can only lead to the worship of a generalized nature-deity with the consequent neglect of the transcendent triune God revealed in the Scriptures. (Jones, 1970:65)

Criticisms have been forthcoming from the Christian left as well, since it is alleged that Teilhard ignored sin and distortion in human affairs. Liberation theologians such as Gustavo Gutierrez have identified Teilhard's thought as overly optimistic. For these critics, it seems as if progress does not meet with genuine suffering or decline for which redemption or liberation is required. Nonetheless, Teilhard's appeal was unmistakable. The best known of his books, *The Phenomenon of Man* became a best seller after it was published posthumously in 1955, almost twenty years after he wrote it. While in China, on an expedition as a scientific advisor to the Geological Survey of China, in the late 1930s, Teilhard became familiar with a vast array of geological forms, mammals, and paleolithic cultures of the Far East.

The broader significance of Teilhard's consilience of science and religion has to do with his fusion, differentiated though it may be, of scientific investigation with a "mystical" vision of reality. This choice would prove to be a decisive feature of many later Catholic efforts to reconcile faith and science. Teilhard's choice was highly influential on widely read American thinkers such as theologian John Haught and ecological thinker and theologian Thomas Berry, for example (see chapter 5). But, it was also decisive for its original alliance with the European (continental) approach to questions of philosophy and metaphysics.

In Europe, by the middle of the twentieth century, philosophy had taken a turn away from speculative and logical form toward what came to be known as phenomenology and hermeneutics. Whereas Anglo-American or "analytical" philosophy maintains a quest for the objective, empirical basis for justified true belief, European philosophy turned toward the manifestations of meaning in the experience of the human subject. Phenomenology, especially as it was introduced by the philosopher Edmund Husserl, came to designate knowledge about the subject, knowledge about how we anticipate, constitute, and communicate our knowledge in the different contexts in which we find ourselves. It is no accident that Catholic approaches to understanding science in relation to religious faith should come to share, simultaneously, a deep interest in the human subject, phenomenology, experience, interpretation, and culture, the very concerns that motivated continental European philosophy. This movement in philosophy not only took place in Europe, but did so in many of its highly Catholic countries, especially France. As such, the Catholic method of reconciling science and religious faith reflects the priorities of twentieth-century European philosophy as well as the traditional Catholic concern for interiority, the inner life of the human subject. This contrasts with the logic-oriented and partly Protestant inspired Anglo-American philosophy that continues to dominate, albeit with less strength than before, in Great Britain, Australia, Canada, and especially the United States.

Teilhard's writings attest to this focus on phenomena, as a philosophical starting point for dealing with science and religion. Even though de Lubac saw his work in apologetical terms, this should not distract us from understanding Teilhard as he saw himself. Teilhard did not see science as a "proof" for the existence of God per se, yet he saw science as offering a means for understanding the world as structured, due to evolution, with just the right elements of purpose and openness that a person of faith would expect to see if one understood God as the world's creator. This is a different way of viewing the question from the way of starting with individual elements. It begins with the whole, with the complex structure of life and beings as they are found in nature.

For philosophical phenomenologists, this way of looking at the world was exercised mainly in terms of social and historical factors. For Teilhard, the starting point is the unified whole in nature. This holist stance stands opposed to a model of scientific investigation that is strictly analytical, in the sense of science practiced with the aim of breaking down wholes into parts in order to understand how nature operates. But, Teilhard carries the phenomenological perspective one step farther still. While most philosophers accept, to some extent, the distinction between objective reality and subjective experience, regardless of their particular interests, Teilhard's understanding of evolution taught him that the natural world is infused with subjectivity and experience.

This view has enjoyed an enduring, sympathetic hearing in Christian thought and elsewhere. Like the process philosopher Alfred North Whitehead, Teilhard believed that matter is not dead, it is alive and directed toward some sort of end. Like the notion of final cause of Thomist and Aristotelian philosophy, Teilhard saw nature in terms of tendency. In Teilhard's case, of course, the tendencies he sees are not static, unchanging tendencies, but rather tendencies that can multiply and adapt over time. The tendencies of material things are nature's way of responding to God. They are the evidence that God directs nature by a spirit that infuses nature. Spirit does not hover over nature as a second reality. For Teilhard, there are not really two realities, and this explains why Teilhard and others argue against the dualist belief of two separate and distinct realities: spirit and nature. It is Teilhard's monism, or belief in one unified reality, that is partly responsible for the suspicions held against him by some within the church. Taken to its extreme, Teilhard's fusion of spirit and matter might appear to overlook both the unique spirit-matter fusion of human persons (the anthropological), and also the uniqueness of Christ (the Christological). According to church teaching, God is revealed through the historical experience of Israel and Jesus Christ, two "events" that very much depend upon the uniqueness of humans to obtain a glimpse of God and on the uniqueness of Jesus' person. But for Teilhard, the union of spirit and matter in the world does not obscure the uniqueness of humanity and Jesus Christ. Rather, Christ is the omega point, the "future" of the *noosphere* (or the level of thinking reality that is networked), and this future is impinging upon the present because of the resurrection. Simply put, Teilhard's contribution to the relationship between Catholicism and science is enormous and his legacy has still to be fully developed.

VATICAN II AND SCIENCE

At Vatican II, the moment of *aggiornamento* was proclaimed. At this time, an alliance with modernity is declared and acted upon in a number of ways, such as the reform of the Catholic liturgy, which until the 1960s was celebrated in Latin, the language of the Church. At Vatican II, it was agreed that the Mass (from the Latin word "mittere," which means the dismissal or last words of the Catholic liturgy) should be celebrated in the vernacular language, the language of the nation or region in which the liturgy takes place.

The opening to the world was also reflected in a new attitude toward science, but this attitude is best accounted for by noticing what is *not* contained in the conciliar documents, rather than what is contained there. Very little substantive comment on science is contained in these documents, yet a significant amount is devoted to issues of technology and ethics. One detects, therefore, a turn away from the theoretical issues on which the Church had passed some critical comments regarding scientifically informed positions perceived to be at odds with church doctrine. At the same time, one detects at Vatican II a turn toward the social implications of science and technology, a concern for the way in which science is embedded in crucial developmental and ethical issues in human culture. Both the social concerns highlighted during the papacies of Paul VI and the ethical issues treated in the various teachings of Pope John Paul II reflect these twin aspects of the implications of science.

In one of the key documents of the Council, *Gaudium et Spes* (the Latin words here mean "Joy and Expectation" and it is subtitled the Church's Pastoral Constitution in the Modern World), science is lauded for its contribution to human culture (n.53). In fact, the concept of progress, itself a key to understanding the thrust of modernity, is very important in the way that the bishops of Vatican II understand what modernity means essentially. There are still warnings against a pure "earthly humanism" and the use of science to support atheism (n.19). But overall, the tone and attitude toward science changes at Vatican II to one of balance, where the values of science are highlighted along with the dangers of science. The document admonishes Christians who would try to downplay the rightful autonomy of science. In 1968, Pope Paul VI took this rapprochement to imply that, as he says,

we are far from the frequently petty and almost always sterile disputes which once gave pleasure to certain minds, inclined as they were to consider the Church, and the advance of human knowledge, as two openly struggling adversaries. (Martini-Bettòlo, 1986:126)

One of the significant themes emerging from the Church's greater concern with the world and political affairs, after Vatican II, is the theme of nuclear and total warfare. It is the scientific basis for such warfare that has driven successive Popes and also bishops and other church bodies to condemn scientific research that directly contributes to harm and destruction. Pope Paul VI, in the same 1968 address, states:

[T]he spectre of most terrible calamities, capable of overwhelming and razing to nothing the whole inhabited earth, rises in fact from the most advanced laboratories of modern physical science. Can we remain silent about such prospects? No matter how great is the responsibility of politicians in this regard, yet the full responsibility of men of science also remains [...] May necessary renunciations be made with courage! Let every measure be taken and every obligation assumed, in order to prevent and avert the manufacture and use of nuclear arms [...] May mankind return to its senses! (Martini-Bettòlo, 1986:127)

It is statements such as these that lie behind the movement later in the 1970s and 1980s by various Catholic bishops, episcopal conferences, and theologians to a more radical peace and disarmament stance. In its acknowledgement of the radical co-opting of science by nuclear states, Pope Paul VI and his successors have in fact tested the limits of the applicability of the just war theory. Arguably, the traditional conduct of war, on which the classical definition of a just war partly depends, no longer applies to some conflicts and other potential conflicts because of the total nature of their conduct. In arguments over war, there is increasingly a "presumption" against conflict. Such is the drift in church teaching on war, which implicates the moral stance of scientists.

JOHN PAUL II AND CONCLUSION

Pope John Paul II has been identified as one whose interpretation of Vatican II has taken the Church in more conservative directions. Whether sympathetic toward him or not, most would agree that he has been a forceful advocate for cultural renewal and Christian identity in increasingly secular social environments. Yet, on relations with science, this has not meant any broad retreat to previous positions, such as the one taken by Pius XII regarding polygenism. The most well-known aspect of Pope John Paul II's assessment of science is his response to a commission of the Pontifical Academy of Sciences into the Galileo crisis. The commission was established by the Academy in 1981. In its final report in 1992, the commission stressed the fact that Galileo did not possess proof for the Copernican hypothesis. This emphasis in the commission's report was criticized heavily as well as the commission's lack of forthrightness regarding the central role of Pope Urban VIII in condemning Galileo. Nevertheless, Pope John Paul II admitted that the condemnation of Galileo centuries earlier was "a hasty and unhappy decision" (John Paul II, 1992: 2).

While some would argue that the Church's conservative positions in bioethics are evidence for some sort of return to a premodern stance, it is nevertheless evident that the Church's positions on ethical issues are borne not so much on the basis of an ignorance of science, but rather on the basis of an interpretation of natural law and social morality. On this question, to which we will return in Chapter 6, the issue is: what is the philosophical position that identifies the natural law as "natural"? The contemporary

Catholic Church does not, by and large, attempt to say whether scientists should probe and attempt to explain nature according to one or other theory in the way that it did with Galileo. The conservatism of John Paul II's papacy is not a set of positions that defies nature and science, but rather a set of positions that seeks to interpret nature and science in a way contrary to the ethical and moral assumptions of many scientists and moral philosophers.

There is one way in which John Paul II's papacy has been marked by a theoretical traditionalism. This is the role of Thomas Aquinas and the emphasis on natural theology in some aspects of John Paul II's teaching. Especially in *Fides et Ratio*, his 1998 encyclical, John Paul II returns to the Thomist tradition as one that is capable of mediating the natural sciences with Christian faith. Unlike *Humani Generis*, this encyclical proposes a positive articulation of the reasonableness of faith:

Just as grace builds on nature and brings it to fulfilment, so faith builds upon and perfects reason. Illumined by faith, reason is set free from the fragility and limitations deriving from the disobedience of sin and finds the strength required to rise to the knowledge of the Triune God. (*Fides et Ratio*, #43)

A central concern for John Paul II has been the rise of evangelical Christianity at the expense of the Catholic Church. In its more fundamentalist forms, evangelicalism has expressed a suspicion of reason and a negative assessment of the significance of secular realms of life and thought. In the context of certain tension between Catholics and evangelicals, particularly in Latin America and Africa, the Church is anxious to buttress its theological reasoning with arguments from tradition. In this case, reason implies a positive assessment of science and a retrieval of Thomism:

Looking unreservedly to truth, the realism of Thomas could recognize the objectivity of truth and produce not merely a philosophy of "what seems to be" but a philosophy of "what is." (*Fides et Ratio*, #44)

There is some evidence therefore for the claim, based on the Church's understanding of the relationship between philosophy and science, that John Paul II is embracing the past. This can be seen in certain statements made in underlining the Church's belief in special divine creative action of the human soul, in distinction to the indirect mediated creation of all other life. As we shall see in the next chapter, this claim, while it marks out a Catholic position in the science–religion dialogue, has not prevented Catholic thinkers of various backgrounds to probe church teaching, whether on this or other matters.

Chapter 5

The Legacy of Vatican II in Cosmology and Biology

INTRODUCTION

In this chapter, we will survey the numerous and complex figures who make up the Catholic scientific, philosophical, and theological scene after the Second Vatican Council. In the name of "aggiornamento," the Second Vatican Council heralded a new chapter in the Catholic Church's relationship with the world. Aggiornamento is commonly interpreted as the watchword for the Catholic Church's turn to the modern world. Typically, this word signals the Church's openness to the findings of history and the social sciences. Another of the catchphrases of Vatican II was the call placed on the Church to actively discern "the signs of the times" in the world, a way of expressing the need for the Church to understand the world in its social and cultural complexities. What this has meant in some regions of the world is a diminished reliance upon a hierarchical form of church organization. However, many would cite evidence of more recent moves to embrace hierarchical authority within the Church. The Church's reluctance to maintain its monarchical image was symbolized in 1978, when Pope John Paul I refused to wear the imperial "tiara" upon being installed as the successor to Pope Paul VI. Pope John Paul II followed suit a few weeks later, and at his death in 2005, commentators noted his simple coffin and lack of earthly possessions mentioned in his testimony as significant indications of a similar spirit of personal humility.

From Vatican II onwards, the exclusively didactic or teaching function of church doctrine came to be explicitly complemented by a stance of listening and solidarity with the world and all its peoples, religions, and traditions. This stance is one of openness, and while it has been interpreted

Roman Catholic bishops from around the world gathered in Saint Peter's Basilica in Rome at Vatican II. The four sessions of the Council marked a turning point in the history of the Church as it recognized the positive elements of life in the modern period with which the Church could assent. (Catholic News Service)

by church conservatives as groundless concessions to the secular world, it is essentially a stance that recognizes the autonomy of various spheres of life, such as politics, business, health care, schools, and science. Vatican II can be seen as a belated yet honest appraisal of the Enlightenment. Interpretations of Vatican II are often wildly different. To exaggerate either the idea that the Church is strictly a teaching institution or a service institution often reflects the perspective of the interpreter, not an objective appraisal of what Vatican II brought about. The ongoing series of interpretations of the Second Vatican Council reveal a great deal of flux and tension amongst church leaders over the theology of the Church.

As noted in the last chapter, Vatican II documents testify to the Church's new attention paid to science, relative to a paucity of prior interest. In Vatican II, one can discern no sudden conversion to the significance of science. But certain texts indicate a self-conscious spirit of openness toward science that was not there previously. *Gaudium et Spes*, the Council's "Pastoral Constitution on the Church in the Modern World," reads thus: "And so mankind substitutes a dynamic and more evolutionary concept of nature for a static one" (Abbott, 1966: *Gaudium et Spes*, #5) and adds later that "historical studies tend to make us view things under the aspects of changeability and evolution" (Abbott, 1966: *Gaudium et Spes*, #54). This indicates at least some degree of rejection of older Aristotelian and scholastic interpretations of nature and history amongst the bishops of that council.

One of the key examples of this shift concerns polygenism, which, as we saw in the last chapter, is the view that there were many first parents, a belief stemming from evolutionary theory's that human beings emerged from hominid and prehominid ancestors in groups or clans. This implication from evolution was denied by Pius XII in *Humani Generis* (1951). But while that encyclical condemned polygenism, there is silence on the topic of polygenism in Vatican II documents as well as subsequent church teaching. In the new Catholic catechism, there is one oblique reference to the issue. In it, the following position of *Humani Generis* is not repeated: "The account of the fall in *Genesis* 3 uses figurative language, but affirms a primeval event, a deed that took place *at the beginning of the history of man*. Revelation gives us the certainty of faith that the whole of human history is marked by the original fault freely committed by our first parents" (CCC, 1994: §390). Again, while the existence of first parents is affirmed, the Church does not claim that these parents are solitary historical figures, which would be contrary to evolutionary theory. The Church here claims the theological truth of sin, entering into human history at an early historical period, an event narrated with reference to Adam and Eve, figures who stand in for all of humanity thence.

In this chapter, attention will be devoted to Catholic responses to the significant developments in scientific theory since the 1960s. Many of these responses have been formulated through the discourse of philosophy or systematic theology. With some exceptions, the more speculative aspects of this thought have not found their way into official church teaching or doctrine. One reason for this is due to the speculative nature of these interpretations of science—the idea that if the Church does not feel a pressing need to speak on an issue (i.e., on issues outside its domain of competence), then it should not do so.

The Catholic Church increasingly takes a stand in favour of dialogue, a stance that presumes the authoritative role various parties might possess over a particular domain while recognizing some overlapping interests. Dialogue is one of four ways to relate science and religion, according to the famous typology of Ian Barbour. In his widely read book, *Religion in an Age of Science*, Barbour relates four basic positions adopted by various churches, theologians, scientists, and philosophers: Conflict, Independence, Dialogue and Integration. Conflict and independence are fairly straightforward positions, even though the motivations and reasons for choosing either position are varied. In conflict, science and religion are opposed to one another in principle, and either one or the other has a more privileged access to truth. According to the independence model, science and religion are separate domains altogether with distinct, identifiable categories that do not impinge on the other domain. Barbour cites several Catholic thinkers as basically sympathetic with the dialogue approach, though some are connected to the independence approach. Teilhard de Chardin and process thought, on the other hand, are examples of integration, according to Barbour. Teilhard's notion of the Omega Point (a discernible goal toward which all of reality is headed), for example, is a scientific category, arising as it does from considerations of evolving cosmic complexity. Yet, it is also a theological category, in consideration of the redemption won by Jesus Christ and his promises to his followers regarding the ultimate destiny of the whole world at its end. In Teilhard's case, as with process theology, there is an explicit theology of nature that is being claimed. A theology of nature, whether expressed in Teilhard's more mystical terms or in more traditional Thomist philosophical categories, is a form of integration. To propose a theology of nature is to suggest that God's creative activity is actually evident or manifest in nature.

One of the central features of the post-Vatican II church is a greater degree of freedom enjoyed by Catholic thinkers, especially theologians, but also philosophers and scientists as well. Alongside the official teaching of the Church, therefore, has grown up a discernable stream of Catholic thought with roots in the tradition combined with ties to various academic disciplines and issues of the modern world. In the context of science and theology, it is true that Catholic thinkers in the area tend to emphasize the role of philosophy. This has helped secure a reputation, cited by Barbour, for an association between the Dialogue position and Catholic thought in the science–religion dialogue. Assumed by Barbour is the emphasis on dialogue that is strongly conveyed by practicing philosophers. As we have already seen, the Catholic strength in philosophy is one reason why Catholic contributions to science–theology dialogue have been both indirect and yet comprehensively framed.

In addition to the emphasis on philosophy, there are three main ideas that Catholic thinkers in the post-Vatican II period have defended. *First is the idea that faith is a strongly personal pursuit.* This does not suggest that faith is a path to be traveled by individuals in isolation from one another. Rather, God's personal character is assumed or claimed in Catholic theology with greater emphasis in comparison with the liberal Protestant approach. Faith is a matter of each one's own personal encounter with God's revelation, disclosed in the three persons of the trinity. Against the idea that faith is an amorphous assembly of spiritual insights constructed on the basis of personal taste or preference, Catholic thought has emphasized the personal encounter with God, revealed primarily through two points of contact: (i) the unique life, death, and resurrection of Jesus Christ and (ii) the Church, which grew out of a collective, personal encounter with Jesus. In thinking about the Church, Catholic thinkers have placed greater emphasis since Vatican II on the role of the Holy Spirit in confronting sin and evil, animating community life, action for justice, and mutual support. This has come by playing down the role of the Church's hierarchical structure and its teaching authority, although such changes in emphasis are by no means universal. What remains clear is the importance of the relationship of church members with God, identified through prayer, sacrament, and the quest to live the moral life.

Second, Catholic thinkers emphasize the unity of reality. Despite the distinctive personal character of faith, Catholic thinkers place considerable emphasis upon a metaphysical form of philosophy that attempts to unite different disciplines or areas of expertise. This second emphasis is not as significant as it once was however. As we have seen in the previous chapter, the dwindling influence of Thomism as the official Catholic metaphysical philosophy has been deeply felt since Vatican II. Nevertheless, the quest for a unifying philosophy for reality, including the ultimate reality of God, continues to be strong, especially in comparison with much of evangelical and liberal Protestant theological thought. As we will see, this quest for unity reverberates in a number of Catholic thinkers over the past fifty years. And, it does so often in a more or less Thomistic vein.

Third, Catholic scientists and philosophers of science continue to be strong advocates of a realist epistemology. The doctrine of realism has two meanings. First, it means that truth can be both discovered and verified, in science and theology, though in different ways. With some exceptions, Catholic thinkers tend to hold out for the reliability of sense perception, the understanding of intellect, and the reflective knowledge of judgment as key cognitional levels that are active in developing truthful knowledge. As we will see, this is especially the case in the thought of Bernard Lonergan.

One aspect of the realist doctrine that stands out in Catholic thought, especially in papal encyclicals and allocutions, is the idea that "truth cannot contradict truth." This affirmation expresses the distinctive, personal nature of religious faith. This means that faith and reason cannot contradict each other. It is a belief, born of experience in both the life of the Church and in science, that the truth of one dimension and the truth of another dimension are complementary. Each dimension complements the other dimension. Of course, this belief stems from the heritage of Thomas Aquinas, whose affirmation of the life of reason was the result of his appropriation or use of the empirical philosophy of Aristotle. Against Bonaventure and other "spiritual" theologians who believed that faith alone was a sufficient basis for theology, Aquinas instead regarded Aristotle's emphasis on nature and empirical knowledge as important supports for the belief that order is created by God. A second aspect of the realist doctrine in Catholic thought is the notion of objectivity. While many Catholic philosophers have been influenced by postmodernism and Continental philosophy, Catholic thinkers still tend to stress the independent, factual, and objective character of knowledge that results from careful, reasoned investigation, whether in science or theology.

A significant example of the Catholic emphasis on objectivity concerns its continuing interpretation of "hypothesis," which as we saw earlier was a key issue in the Galileo controversy. In 1996, Pope John Paul II issued a statement on evolution, which reads: "Today, more than a half-century after the appearance of that encyclical, new knowledge has led us to realize that the theory of evolution is *no longer a mere hypothesis.*" Rather than appear to remain skeptical of evolutionary theory in biology, the Pope chose language that would indicate, as directly as possible, that the Church did not wish to repeat history by deeming speculative or "hypothetical" what is, in fact, verified. Another (second) translation of this speech interprets the last part of the French sentence (" ... à reconnaître dans la théorie de l'évolution plus qu'une hypothèse.") to read as follows: " ... some new findings lead us toward the recognition of evolution as *more than an hypothesis*" (Pope John Paul II, 1996, para 4). The first version is authoritative and contains stronger connotations that suggest the truth involved in the verification of the theory of evolution. The second version is not quite as confident in the explanatory power of evolution.

There is actually a third version of the statement which appeared originally in the Vatican newspaper *L'Osservatore Romano* (#1464, October 30, 1996). This third version read as follows: " ... some new findings lead us toward the recognition of more than one hypothesis within the theory of evolution" (Eternal Word Television Network, website footnote translation).[1] As the Jesuits in Science web newsletter suggests of this interpretation, this translation "seems to be a significant misunderstanding of the French original" (Jesuits in Science, 1997).[2]

Some conservative Catholics in the United States, citing the (second) French translation, have interpreted the Pope's statement in ways that benefit the skeptical approach toward evolutionary theory, as if this skepticism is the meaning that John Paul II intended. This conservative reaction takes advantage of the confusion in translation in order to cast doubt on Catholic teaching concerning evolution. The original text that reads "no longer a mere hypothesis" is a logical choice for the meaning of the Pope's sentence, since it so exactly mirrors the language used to condemn Galileo's defense of the heliocentric universe, a "mere" hypothesis. But the conservative antievolutionary reaction also attempts to build on the way that the Pope speaks about evolution elsewhere in the same speech.

For instance, the Pope also states: "rather than speaking about the theory of evolution, it is more accurate to speak of the theories of evolution. The use of the plural is required here—in part because of the diversity of explanations regarding the mechanism of evolution, and in part because of the diversity of philosophies involved" (# 4). What this diversity implies, for conservative critics of evolutionary theory, is that evolution, being conceptualized in more than one hypothesis, has not been clearly verified. This view is also partially supported by the philosophical language that the Pope uses in this speech about the uniqueness of the human species, which represents "ontological discontinuity" in the universe. This is a discontinuity between the human level and the physical and chemical levels that support human life. What such controversies as this one over papal language on evolution show is that there continues to exist many profoundly methodological, philosophical, and theological ambiguities concerning the Church's relationship with science. The rest of this chapter turns to ways that various Catholic thinkers have handled such ambiguities since the 1960s. One cannot escape the conclusion that there is a broad diversity of views adopted by Catholic theologians and scientists on the relationship between the Christian faith and science.

HANS KÜNG

One of the first Catholic theologians to see the importance of drawing connections between theology and science in the aftermath of Vatican II's embrace of the modern world is Hans Küng, a German theologian from Tübingen, a well-known German centre of theology. A father or *"peritus"* at the Council, Küng caused a controversy in the 1970s and 1980s by criticizing the doctrine of infallibility and the papal encyclical *Humanae Vitae*, which argues that artificial birth control methods are immoral forms of family planning. Undoubtedly, he is one of the first Catholic theologians to advocate explicitly for a theological dialogue with the human sciences in order to ensure that theological statements retain the rational aim of plausibility. His contribution was widely seen at the time to be in the *avant garde*, part of the movement leading Catholics toward full aggiornamento, full acknowledgement of modern life. His criticism of the ways in which Vatican II is interpreted by the Church led to his estrangement from the Church, although he has since met with his old archrival Cardinal Ratzinger since the latter became Pope Benedict XVI. In line with the spirit of modern historical and scientific criticism of traditional belief, Küng was one of the first theologians to express his skepticism over some of the claims contained in the New Testament concerning miracles, and the bodily resurrection of Jesus. For example, on the resurrection, he writes: "It was admittedly not a historical event (verifiable by means of historical research), but it was certainly (for faith) a real event" (Küng, 1976:351).

Küng is also known in terms of the connection between science and theology for his use of Thomas Kuhn's theory of paradigms. Kuhn was a philosopher of science who advocated the view that science does not proceed gradually. Instead, "science enjoys periods of stable growth punctuated by revisionary revolutions" (Stanford Encyclopedia of Philosophy). For Kuhn, science is a series of incommensurable systems of thought. And, in his widely read book *Does God Exist?* Küng asks the question, "Are there 'scientific revolutions' also in philosophy and theology?

As a natural scientist, Kuhn did not deal with this question. It is however, scarcely possible to deny it" (Küng, 1981:111). In a 1983 conference and follow-up book, Küng lays out five stages or paradigms in Christian theology that parallel the ways in which science proceeds in periods of "normal science" interrupted by brief, revolutionary periods in which the entire conceptual framework of the previous period of normal science is dismantled. According to Kuhn, there are five major periods of Christian theology as a result of this categorization: Greek Alexandrian, Latin Augustinian, Medieval Thomistic, Reformation, and Modern-Critical. In summary, Hans Küng does not draw on science per se. He does advocate the view that a contemporary historical understanding of science has direct implications upon the way in which we understand the history of theology. Küng provides legitimacy for the idea that changes can be made in theology, by virtue of the fact that changes occur in science. The widespread, universal acknowledgement of science's objectivity therefore suggests to Küng that if radically paradigms of knowledge exist in science, then the same holds in theology.

WILLIAM STOEGER AND GEORGE COYNE

William Stoeger and George Coyne are Jesuit priests and astronomers. Coyne is the director of the Vatican Observatory, located in Castel Gandolfo, Italy, where the Pope's summer palace is situated. Stoeger is a member of the Jesuits' American observatory just outside Tucson, Arizona, where he splits his time between teaching duties there and research activity in both Arizona and Italy. Coyne has been involved in developing an international reputation in scientific circles for the work of the Vatican Observatory, and is an expert in interacting binary star systems.

Stoeger has written a large number of articles, many of which are contained in a publication series known as the "Scientific Perspectives on Divine Action" cosponsored by the Vatican Observatory and the Center for Theology and the Natural Sciences, based in Berkeley, CA. He has been one of the most assiduous proponents of a Catholic approach to science–theology dialogue through the promotion of a philosophical mediation. A philosophical mediation of the dialogue does not, in itself, make this kind of approach Catholic. Yet Catholic doctrinal statements and Catholic scholars in this field do tend to emphasize the role of philosophy relative to other Christians for reasons based in philosophy and natural theology.

Stoeger's own view of philosophy is most pronounced in relation to cosmology, as when he writes:

[...] physics and cosmology do not presuppose the conclusions of other disciplines—as does biology relying on chemistry and physics, and chemistry relying on physics. When we step back from physics and cosmology to justify the assumptions and presuppositions we employ in pursuing them, we have nowhere to go, except to some sort of philosophical reflection. (Stoeger, 1988:227)

This is one of the most straightforward ways of putting the problem of ultimate meaning—as a philosophical problem emerging directly from a consideration of scientific limits by scientists themselves. The most extraordinary example of this kind of overlap between science and philosophy is in the discussions over the anthropic principle. The anthropic principle states that the universe appears to be so finely tuned to produce carbon-based life, the idea that thinking creatures like human beings emerged is because the necessary natural elements existed from the beginning, predisposing an outcome like the emergence of a thinking species.

As an editor and regular contributor to the Vatican Observatory/Center for Theology and the Natural Sciences (CTNS) research volume series on "Scientific Perspectives on Divine Action," Stoeger has been a consistent advocate for what has been dubbed the double causality model of divine action. This model of divine action can be confusing to understand. Like many others involved in this research project, Stoeger holds that God's action can be distinguished in two basic forms: universal or general divine action (GDA) and special divine action (SDA). Usually, GDA is associated with the action of creating the world, both in the beginning and continually. SDA is associated with the person and works of Jesus Christ, the experience of God through the work of the Holy Spirit and other revelatory or one-time only events like miracles. Where this becomes confusing is whether or not God's special action includes specific actions in and through the natural world. Some other participants in this field think that there exists a unique set of natural processes which demonstrate God's continuing special action in sustaining the universe. This is the search for a "causal joint" between God's causal action and the universe. The anthropic principle is frequently cited in this context to suggest a causal joint at the macrocosmic scale, the purposeful fine-tuning of cosmic features so as to make the universe habitable for human creatures. What advocates of the causal joint also seek is an indeterministic picture of the universe: at various levels of nature, there exists an inherently open, undetermined arena in which God can continue to exercise creative action in causing or persuading free entities and agents in certain directions.

Against a simplistic search for a causal joint, Stoeger believes that the debate over whether nature is deterministic or indeterministic is unimportant for understanding how God acts in the world. The search for a "causal joint," especially the approaches led by Nancey Murphy and Robert Russell, founder of the Centre for Theology and the Natural Sciences in California, is in line with the Protestant tradition of citing direct evidence in nature for God. This form of natural theology was prominent in the seventeenth-century natural philosophy of Newton and Boyle. Russell, on the other hand, sees the inherent openness of nature at the quantum level as the level of reality where God continually creates from the "bottom-up." At this level moreover, there is no need to worry that such action constitutes a view that God is violating natural laws. This is because at the quantum level, there is inherent indeterminacy. There are no laws about which God's actions would be conceived as violations.

In contrast, Stoeger's approach is recognizably Catholic for its promotion of analogy. God's being can be inferred to be *like* the intelligibility of the world and from experiences of goodness, beauty, and truth or order. Stoeger's approach resembles the view that God's continuing creative action can be mediated through secondary causes without becoming negated as ultimately divine action. Moreover, it is not necessary to search for further knowledge about nature and natural laws in order to see how divine action is unfolding. Yet, Stoeger also suggests that aspects of natural reality are entirely consistent with the kind of world one would expect to find given that it is fashioned from a creator: "there really is a directionality manifest at the level of the sciences in the evolutionary process as a whole" (Stoeger, 1998:165). In summary, Stoeger's approach is probably one of the most scientifically informed and nuanced Catholic approaches to the dialogue, especially in dialogue with physics and cosmology. Stoeger is comfortable with affirming divine action while at the same time suggesting that science places genuine constraints on how we can talk about divine action.

KENNETH MILLER

Kenneth Miller is a molecular biologist from Brown University and the coauthor, with Joseph Levine of one of the most popular Biology texts used in North America, *Biology*, which is published in numerous editions. Partly due to his visibility as a textbook author, Miller has become involved in some of the high-profile debates over evolutionary theory in science curricula. Miller himself is a strong defender of the theory of evolution and its inclusion in the Biology curriculum, although he disagrees with the materialist interpretation given to evolution. He also points out that there are various theories and versions of the theory of evolution within what is usually referred to as "the" theory of evolution. This is one of a number of key points made in his popular book on the interaction of science and religion, *Finding Darwin's God*.

In this book, Miller defends evolution as consonant with a religious framework, although most of the book is given over to quite detailed descriptions of evolution, defined as "descent with modification." Evolution

Professor Kenneth Miller of Brown University, a Catholic and author of the book *Finding Darwin's God*. Miller has strenuously defended Darwin's theory of evolution against attacks by young earth creationists and Intelligent Design theorists. In 2005, he appeared as a star witness in the Dover trial in Pennsylvania where a school board's plans to allow the teaching of Intelligent Design theory was challenged and defeated. (Courtesy of Brown University)

is far and away a better theory than alternative theories such as "Intelligent Design" in accounting for complex cells, organs, and living organisms. At the end of the book, Miller endorses a minimalist view of God from the vantage point of evolution, borrowing from the final sentence of *The Origin of Species* by Charles Darwin himself:

There is grandeur in this view of life; with its several powers having been originally breathed by the Creator into a few forms or into one; and that, whilst this planet has gone cycling on according to the fixed law of gravity, from so simple a beginning endless forms most wonderful and most beautiful have been, and are being evolved. (Darwin, 1956:560; Miller, 1999:292)

Miller adopts the most prevalent position taken by Christian biologists with this view of divine action being limited to an initial creative act. God was present in the original structuring of carbon-based life, which gradually differentiated into the millions of species and genera that we know today, including those countless species which have become extinct. Although he publicly identifies himself as a Catholic, Miller restricts himself in his writings to scientific issues, instead of doing theology. He bases his criticisms of Intelligent Design theorists and Young-Earth creationists on scientific research. This is itself a mark of a Catholic approach actually. Catholic scientists tend to be more comfortable with the independence of science from religion.

However, one of the strongest points Miller makes is philosophical. He is unwavering in his conviction that when belief in God arises from some gap in human scientific knowledge, it is a flimsy sort of belief. Taking the example of why a plant produces flowers, for instance, Miller chastises the parish priest from his childhood, Father Murphy. This priest pointed out in the mid-1950s that no scientist had yet explained how a plant makes flowers. He concluded that the scientists would not be able to explain this phenomenon. Therefore, only God could be credited with such a creation. But, Miller recalls how, while attending a scientific conference several decades later, he heard a detailed account of how plants do make flowers, in part because the four parts of a flower—the sepals, petals, stamens and pistils—are actually modified leaves. For Miller, this example, though simple, carries great lessons for understanding the purported theological implications of Intelligent Design theory.

One of Miller's strongest arguments is his demonstration that Intelligent Design theory relies upon the idea that God or an external intelligent agent must intervene in the creation of certain biological structures to render these structures "irreducibly complex." The problem with such a designer of nature is that it implies a purposeful deceiver. For example, take the animal species on the Galapagos Islands, just west of Ecuador in South

America, and the Cape Verde islands located just west of northwest Africa. These are island archipelagos with species that resemble species found on the respective nearby continents. If Intelligent Design theory were correct, then each individual species is independently created by God. But if that were the case, then, as Miller suggests, "one would have to believe that it was also the designer's choice to mislead—by producing sequences of organisms that mimic evolution so precisely that generations of biologists would be sure to misrepresent them" (Miller, 1999:94).

Obviously, one problem for Intelligent Design theorists is that their theory commits them to saying in principle that if the theory of evolution cannot account for the emergence of species in macroevolution, then every living species, the whole "stunning diversity of life" as Miller terms it, is specially created by God. As Miller says, such a theory contradicts the best available scientific evidence. As Miller points out, why else would human embryos form a yolk sac during the early stages of development as birds and reptiles do, since they are placental animals, which means they draw nutrients directly from their mothers? What is at work here, biologically speaking, is not intelligent design but evolutionary ancestry: humans, like other mammals are descendents from earlier reptile-like animals. More seriously perhaps is the fact that Intelligent Design theory contradicts, ultimately, the theological understanding of God who bestows independent powers to the natural world according to many long-standing interpretations of the doctrine of creation. Miller also cites underlying moral issues in the evolution-creation debates. Intelligent Design and Young-earth creationists see the decline in morality as the most important effect of a worldview in which materialism triumphs over God, who has been pronounced dead. But Miller believes that moral behaviour does not depend on a special, divinely unique creation of species.

Does Miller advocate any view of nature that might support the existence of God? Although not explicitly, Miller nevertheless suggests that the anthropic principle lends credibility for arguments in favor of God's existence. Such a view is not the same as the view of God as an active intervening agent according to the theory of Intelligent Design. Intelligent Design theory renders God active as the cause of an ongoing series of miracles in order to account for the macroevolution of species in history. Miller emphasizes the basic building blocks of the universe, physical building blocks such as the numeric specificity in gravity, without which the element of carbon and hence life itself, could not have emerged. In short, Miller finds many reasons to doubt the view that God is as active as antievolutionists claim. But, Miller does agree that science is ultimately limited in its capacity to understand why the universe is the kind of universe it is.

Profesor Michael Behe, a Catholic and biochemist whose book *Darwin's Black Box* has been one of the key texts in defense of the claims of Intelligent Design theory. Behe emphasizes "irreducible complexity" in characterizing certain biochemical processes and entities. His claims have been discounted by other scientists for various reasons, including his lack of publications in peer-reviewed biochemistry journals. (AP Photo)

MICHAEL BEHE

Following earlier Catholic skepticism toward evolution, biochemist Michael Behe is one of the few Catholics at the forefront of the Intelligent Design movement. Along with well-known mathematician and philosopher William Dembski, Behe has been at the forefront of Intelligent Design. Behe believes that the nature of a cell, an important unit of biological study, is "irreducibly complex." That is, the cell is a kind of "black box," a level of living reality that defies causal explanation in terms of the constituent parts that make up the cell. One early exponent of such an approach in biology was the Catholic scientist and philosopher, Lecomte du Noüy, a Belgian who wrote in the 1940s and 1950s. Du Noüy claimed that the first living cell could not have come about from natural forces only (McMullin, 1988:68). He supported such references to the miraculous by citing, for instance, the emergence of green algae from blue algae:

> ...a series of unknown phenomena ended in the appearance of very elementary algae which still exist today, the Cyanophyceae or blue algae. In some of these, the marvelous chlorophyll is not yet present. Their pigment is phycocyanin. These plants resemble the bacteria by their tubular or spherical form and by their asexual production. They perfect themselves (?) and one day there is a great advance: the green algae invade the waters with, at last, the hope and the possibility of a conceivable evolution. They have a nucleus—which is a kind of miracle—and it seems that they inaugurate sexual production—another miracle. Do the green algae with a cellular structure and a nucleus really derive from the blue algae? We cannot affirm it. At any rate the difference between the two is tremendous and the mechanisms of transition again inconceivable. (du Noüy, 1947:51–52)

There are two noticeable features of this statement. The first is the use of the word "miracle." However, in this context, it is meant metaphorically, not as an actual divine intervention which accounts for the sudden appearance of the green algae. The second thing to notice is that evolution is not denied. In fact, du Noüy continues in his book, *Human Destiny*, to state categorically that one cannot deny the theory of evolution in general. What is being suggested then? Well, as Behe also claims, evolution cannot proceed on its own through "natural selection and mutation." It requires guidance acting over it, what the materialist philosopher Daniel Dennett disparagingly calls "skyhooks." For Dennett, these are fictional cranes from above that would intervene to facilitate the evolution from one species to another. These skyhooks literally lift some core features of a more primitive species and provide it with greater functionality at a higher level to produce a more complex species. But no such mechanisms exist according to Dennett, because none have been discovered or ever will be.

But for Behe, one example of intelligent agency that is non-Darwinian is vision, the capacity to see in higher animals. For him, such capacities are like "black boxes," a complete mystery for Darwinian researchers. This is discussed in Behe's book, entitled *Darwin's Black Box*. On a purely Darwinian view of it, vision would have to evolve through slow, steady improvements. But, the problem for Darwinian evolution, according to Behe, is that the eye is a highly specialized machine that either works "as a whole or not at all." Thus, it could not have evolved incrementally, because there is no survival value for any creature possessing parts of an eye, such as a retina without a lens or vice versa. With only parts of an eye, one cannot see. It makes no sense then to say that the eye evolved as Richard Dawkins and other "neo-Darwinists" claim, given that survival is, according to these neo-Darwinists, the sole reason governing the process of evolution. What survival value does part of an eye have, unless there already exists a plan for the formation of the entire eye? Yet, such an anticipation or evolutionary foresight is ruled out by the deterministic, random character of Darwinian natural selection. This whole problem, by extension, throws doubt on the entire evolutionary path first sketched by Darwin himself on the emergence of vision in higher animal forms. According to Darwinian analysis, what was merely a simple group of pigmented cells serving as a light-sensitive spot in early creatures serves as the evolutionary origins of vision in higher animals.

One of the most compelling aspects of Behe's argument is his clarification of Intelligent Design theory. He stresses that this theory is not one which implies perfection of systems created by an Intelligent Designer (God). This is an important revision of the classic arguments made for Intelligent Design by the likes of William Paley. Instead, analogous to the concept of "built-in obsolescence" in engineering, the inference to intelligent design in natural systems is an inference to a system that had to emerge "irreducibly complex" (Behe, 1998:223). That is, Intelligent Design is an inference made for systems that are made up of different component parts, each one of which is necessary for the functioning of the whole system.

JEAN LADRIÈRE

Born in 1921, Jean Ladrière is a Belgian philosopher of science who has written about the meaning of science in relation to epistemology and religion for many years. Of all the thinkers treated in this chapter, Ladrière is perhaps the most abstract. His philosophical background is a mixture of Ludwig Wittgenstein-inspired linguistic philosophy and the European continental philosophical tradition. Ladrière works within the traditions established in European philosophy by Immanuel Kant and Edmund Husserl. Kant is best known for arguing against the possibility of metaphysical proofs for the existence of God, and Husserl is one of the founders of the phenomenological tradition in philosophy, as mentioned in chapter 4.

However, unlike Husserl, Ladrière believes that phenomenological philosophy can be deployed within science itself in order to better understand scientific reason and the sense of objectivity that science provides. Philosophy, instead of being a discipline that mediates from outside science and theology, acting as a go-between, is actually best utilized in understanding scientific and theological *language*. Philosophy is implied in a similar way in both disciplines as a result. In his career, Ladrière devotes most of his attention to science. One cannot do science or theology without in some way speaking and acting philosophically according to Ladrière. Reason lies at the heart of science as part of human experience, not beyond human experience. To interpret an aspect of reality is itself an event in the universe, and as such, is the object of phenomenological study, as he states:

> ...phenomenology can be defined...as the self-understanding of experience... Phenomenology places itself in a point of view from which it is able, at the same time, to show how science is possible and how it enters into the reality of history...Phenomenology analyses the relationship between man and the cosmic world, trying to describe how the world gives itself to existence, in perception and action. (Ladrière, 1999:225)

Ladrière's preoccupation with language is largely uncontested within the science–religion dialogue, simply because there are few thinkers who really understand the groundbreaking work of Ludwig Wittgenstein and others who follow Wittgenstein's line of thinking. Ludwig Wittgenstein was a German who worked in Britain for a large part of his life. He famously advocated a view of language in which linguistic meaning is differentiated according to the different contexts in which it is used. There is no one ideal for any particular word according to him. One can appreciate what this view of language might mean in the science–religion dialogue— if each discipline or specialty possesses its own world of meaning independent of other disciplines and specialties, then science and religion are entirely separate pursuits, each with their own language. The realities of the world and God are language realities that orient the disciplines, but whose objectivity is impossible to affirm, because it is obscured by the inherent lack of clear meaning for words. Wittgenstein describes this situation as one of different "language games." There is no ideal meaning in a sentence or an assertion. Each and every sentence requires an interpretation from within the particular context in which it is read. Wittgenstein is thus usually interpreted as an antirealist: the objects referred to in the language do not necessarily exist independently of the linguistic context in which they are used. Language constrains knowing.

Ladrière takes Wittgenstein's view seriously, but he sees language in a somewhat different light. In examining sentences or statements in a discipline, Ladrière identifies three dimensions of the meaning of a sentence. One of these is the "illocutionary" force of a sentence and Ladriere applies it to an understanding of religious language: "Religious language is not a speculative language, like that of theoretical cosmology or even like that of metaphysics. It is a language which has a radical self-implicative character..." (Ladrière, 1987:6). The distinction can sometimes be obscured because expressions of faith may "use cosmological representations... of the mythical type" (Ladrière, 1972:149). Religious language is a kind of performative utterance. For Ladrière this does not mean that religious language simply expresses a state of mind, since other forms of language can do that. True, religious language is referential, and the creed is an example of such language, made up of meaningful "I believe" or testimonial statements. For Ladrière, "the relation of an assertion to its truth is a function of the way in which it produces its meaning" (Ladrière, 1987:10). Contrary to psychological reductionists, meaning for Ladrière is not a simple case of wish fulfillment or projection. So, in this sense, he is responding to both the antirealist character of Wittgesteinian philosophy while encouraging a more integrated approach to personal fulfillment in opposition to antireligious interpretations of Wittgenstein. His perspective on religious meaning certainly differs from those of Freudian psychology. And, his view of the authenticity of faith within the grounded language of a religious tradition also counters critiques of institutional religion as "illusion."

ERNAN McMULLIN

In an introduction to a book of lectures and discussion delivered at a conference on Religion, Science and the Search for Wisdom, David Byers notes that "Father McMullin is perhaps the best-known Catholic writer and lecturer on the history and philosophy of science" (Byers, 1987:4). Born in Ireland, McMullin was ordained a Catholic priest and later, after studying in the philosophy of physics in the 1950s, he moved to the United States, taking a position as professor of History and Philosophy of Science at the University of Notre Dame in Indiana.

McMullin is best known for introducing two concepts, one in the science–theology dialogue directly, and another one in the philosophy of science. First, McMullin has proposed the concept of "consonance," a view that science and theology need to remain autonomous intellectual pursuits while nevertheless agreeing to fashion inquiries in such a way that the two disciplines should not fall into conflict. McMullin is not alone in advocating this concept, but he is probably the best well known Catholic associated with this position. In the famous typology for science and religion, Ian Barbour states that there are basically four positions to which almost all thinkers in the dialogue fit regarding the nature of the relationship between science and religion. The four positions are: conflict, independence, dialogue and integration, and this typology is widely accepted, even though some have attempted to revise it.

Professor Ernan McMullin of the University of Notre Dame is one of the world's leading Catholic historians and philosophers of science. His position of "consonance between science and religion" as well as his theory of "retroduction" in explaining how scientists define scientific knowledge are the two critical components of his Catholic approach to philosophy. (Sijmen Hendriks Fotografie)

McMullin falls into the independence position according to Barbour, and McMullin's own profession of consonance between the disciplines seems to confirm Barbour's judgment that McMullin thinks of theology and science as two distinctly different enterprises in knowledge. Yet, consonance aims to say something more than the fact that the two disciplines are independent of each other. For disciplines to be in consonance, they need to be coherent with one another too, even though they are dissimilar in contents. Structurally, many view the science–theology dialogue as two disciplines that function according to similar epistemologies, with science yielding verified hypotheses while theology yields core doctrines. This similarity in structure has been taken to be a common position in epistemology called "critical realism." But McMullin sees even the structure of knowing in the two disciplines as distinct kinds of critical realism: "arguments [. . .] for an appropriate doctrine of critical realism in theology [. . .] would have to be of a kind very different from those relied on in regard to natural science" (McMullin, 1998:23).

The second major contribution to the dialogue is through McMullin's introduction of a term "retroduction," which is a term describing how and why scientific theories successfully account for aspects of the natural world. For McMullin, who is a scientific realist, theories can be progressively verified, even though most of the time they may not achieve a hundred percent degree of certitude. McMullin has spent much of his career as a philosopher of science clarifying and disputing the approach taken by sociologists, philosophers, and historians such as Thomas Kuhn, Sandra Harding, and Steven Shapin. For these thinkers and their intellectual cousins, scientific theories are always constrained by particular data that are theoretically interpreted, not pure. Historical contexts shape the theorist's inquiry and the set of values that animate particular scientific approaches according to the sociological perspective in the philosophy of science. For this school of thought (called the Edinburgh school in the sociology of science), particular theories do not refer to actual physical entities, because we do not know for certain whether those theories will endure, just as Newtonian notions of space and time, once held to be the bedrock of physics, were overthrown by Einstein's theories of general and special relativity.

For McMullin, there is an important distinction to make between those social values that animate scientific work and those values that are uniquely cognitive, or directed toward verifying a theory for its potential success in explaining the cause or causes of natural events. Retroduction refers to this twofold presence of values in science. First, scientists discover a theory by discovering a natural law or gaining an insight into a previously unexpected occurrence. (For instance, the noted French scientist, Louis Pasteur discovered vaccination almost by chance when some chickens were injected with a culture that had aged, which thus boosted the chickens' immune system by the time a second, new and therefore stronger virus was injected with the (then surprising) result that no chickens died when injected a second time.) Second, scientists employ various criteria in order to verify this theory to see if it holds up under various tests or experiments in which variables are ruled out to see if the constant that is predicted by the theory holds true—is it responsible for causing what the theory says it does? In the case of Pasteur's theory of vaccination, there have been millions of subsequent vaccinations carried out on many species, including humans, which successfully demonstrate the power of the original idea that injecting an animal with a culture (that is constituted by a form of the virus being vaccinated against) will stimulate the immune system to resist future higher exposures to the virus.

The significance of retroduction, for McMullin, is that it explains the structure of scientific knowledge without dismissing the contingencies of different historical contexts. A confidence in human knowing through scientific activity means that the human imagination is powerfully successful in discovering and verifying what is true. Since religion and theology also rely upon the imagination, we have additional corroboration of the realistic character of theological rationality. In science, the imagination is important for understanding unobservable entities. The same can then be said for theology. There is, in principle, no difference, whether the mind is involved in understanding physical entities or God. Our imagination is a critical element in coming to discover and in coming to verify. We are therefore justified to say that in science, nature is essentially trustworthy. As the philosopher Michael Polanyi says, science is like other forms of human knowing because it needs to trust nature in order to know nature: it exhibits a "fiduciary" outlook (Polanyi, 1964:299). Therein lies the connection between retroduction and religious faith. We can be reasonably confident that statements and doctrines associated with religious faith are roughly analogous to hypotheses that have been verified in scientific inquiry. We can approximately verify knowledge of unobservable entities, whether in science or in theology.

In summary, McMullin focuses his efforts on key historical episodes to demonstrate why it is possible to achieve consonance between science and religion, and in setting out this task, he eschews the more traditional approaches of neo-Thomist scholasticism. He distrusts traditional metaphysics, because of their lack of tangible contact with science as it is practiced today. Scientific practice can be understood comprehensively without such a metaphysics while still being explanatory and realist. The theory of retroduction provides the necessary support for this historical approach and at the same time, it can be a rough analogy to the way that theology affirms the reality of God.

JOHN HAUGHT

One theologian who has attempted to revise Barbour's fourfold typology is John Haught. Haught is the most well known Catholic exponent of process theology, along with Joseph Bracken (see below). In his book, *Science and Religion: From Conflict to Conversation*, Haught outlines a similar set of four positions in the dialogue: conflict, contrast, contact, and confirmation (Haught, 1995:9).

Process philosophy deeply affects Haught's interpretation of Catholic doctrine, and it seems evident that Haught's Catholicism has also impacted his interpretation of process philosophy. Process philosophy is a tradition of thought that has its origins in the thought of Alfred North Whitehead. In it, process philosophers stress the importance of metaphysics, in particular a metaphysical view of nature which sees "becoming" over the classical and static category of "Being." Process thinkers emphasize the primacy of events over static matter, the interconnectedness of events and the organismic kind of organization in the universe and all its entities, rather than the metaphor of the machine (Barbour, 1997:284). God's reality is also described with an emphasis on change and flux. God is dynamic not static. God is dipolar, conceptualized by Whitehead as a being who is both "primordial" and "consequent," analogous to all "actual entities" and affected by what goes on in the world (Whitehead, 1978:345f). In short, Whitehead's philosophy of the world and God is deeply informed by evolution.

Haught incorporates process thinking into his theological dialogue with science, but he has gone beyond the metaphysical language of process thought to tackle issues at the forefront of theological engagement with evolution, an effort that has resulted in numerous books and articles. Since 2000, in what may be a growing reflection upon his Catholic outlook, Haught has turned to a sustained critique of philosophical naturalism. This is ironic perhaps, given that he is a process thinker. Recall that process philosophy is reputed to be deeply informed by science and evolution in particular and so process philosophers are not usually hostile to naturalism. Haught's problem with naturalism (the view that physical nature is equivalent to reality, with no supernatural or inherently mysterious character to it) is that naturalism is *un*scientific: "Naturalism *assumes* that a kind of cognition that suppresses personal knowing, common sense and teleological considerations can be trusted to put us in touch with what is really real . . . naturalism cannot actually demonstrate that nature is devoid of purpose. Rather it must decree *dogmatically* that purpose cannot be a real aspect of nature . . . " (Haught, 2006:121).

Inspired by Teilhard de Chardin, Haught has been one of the most forceful and articulate proponents of the view that evolutionary biology includes human and cosmic meaning as distinct, intended elements of the universe. Obviously however, Haught does not sympathize with Intelligent Design or other anti-Darwinian movements in trying to correct reductionism amongst evolutionary biologists. In fact, Haught criticizes certain aspects of traditional theology, especially the doctrine of original sin, in which evil is understood as an exclusively anthropocentric reality. That is, according to Haught, original sin is the claim that the condition of sin is a human condition and only a human condition. Rather, according to Haught, sin, evil, and death, while interrelated, are widespread throughout the universe, to which the scope of God's redemptive action is a grander response. Haught infers to this cosmic scope of sin and redemption, in part because the universe is unfolding. Similar to those who interpret sin as finitude (see chapter 6), Haught sees complicity with evil as a refusal to participate with creation rather than a break with a perfectly harmonious original world. Haught depicts God as one whose activity in the world is a "cosmic adventure," one that cannot be simplified as a story of a single historical failure that allowed for sin to enter the world followed by Christ's single atoning sacrifice on the cross for future redemption.

BERNARD LONERGAN

Among Catholic thinkers of the twentieth century, Bernard Lonergan stands out as one of the most precise and complex philosophers and theologians who has incorporated the natural sciences into a coherent yet massive philosophical system. At the time of his writing of *Insight*, his magnum opus, in 1957, *Time* magazine hailed Lonergan as one of the world's leading philosophers. Lonergan was born in Buckingham, Québec, Canada, in 1904 and specialized in mathematics before becoming a Jesuit priest, whence studies in philosophy and theology took him to London and Rome. He is best known for developing a theory of human cognition, which he believes operates at four distinct levels: attention, understanding, judgment, and decision. This theory is developed in his book *Insight*. These four levels were later supplemented by Lonergan and others who have added a psychic level as well as a transcendental level of human loving to the cognitional picture.

Bernard Lonergan, one of the twentieth century's most prominent Catholic theologians, developed a methodological framework for theology that parallels the-accepted model of scientific method. His earlier work in epistemology is a theory of cognition that harmonizes scientific with religious knowledge. (Lonergan Research Institute)

The significance of this theory is twofold. First, it corrects what Lonergan and other "transcendental Thomists" saw as the neo-Thomist overemphasis on categorical metaphysics. The problem that Lonergan and like-minded Catholic philosophers have addressed is the tendency amongst traditional Catholic thinkers to use Aristotelian categories of substance and accident. For Lonergan and other transcendental Thomists, these categories are insufficient to account for the way nature behaves dynamically at all levels. Lonergan saw early on from his study of Aquinas' work that metaphysics would be unhelpful for philosophers if it was not based on an understanding of how people understand. That this issue has been so controversial since Descartes does not necessarily help Lonergan convey the urgency of his claims. Lonergan's focus on how human beings think has persuaded some that he was merely interested in the experience of knowing. But a glance at his writing reveals a deep concern with the kinds of objects that we know when we think. At the forefront of his thinking is the effort to explain why the natural sciences are successful, not as a unique branch of human knowledge, but simply as an outstanding instance of what happens when human beings put their minds to paying attention and gaining insight into a patterned reality.

Science in Lonergan's philosophy is a branch of knowledge that elicits insights successfully because historically, science freed itself from an overweening philosophical worldview that stifled scientific curiosity. But, from science, the most interesting thing is not any one particular theory or data, but rather the presence of insight itself. Human insight is a phenomenon that needs to be understood. From this experience of insight, Lonergan develops a way of describing God by elaborating on Aristotle's notion of a pure, unmoved mover or being. Lonergan's understanding of God is of an "unrestricted act of understanding." In this vein, knowledge of God can be characterized in contrast to the limited nature of scientific knowledge. On the one hand, scientific knowledge is reproducible, and since there is an invariant structure of human cognition, all knowledge of the universe is verifiable, beginning with questions that arise on whether insights and discoveries are actually true. On the other hand, knowledge of God seems to depend on one-time events in which God communicates to humanity certain unique things, or revelations. However, there is one crucial connection between scientific knowledge and knowledge of God. This connection is transcendence or self-transcendence. Scientific knowledge is a form of self-transcendence that becomes fulfilled in religious knowledge.

Transcendence is a subjective, intentional movement to go beyond one's present knowledge by raising further questions: "Man wants to understand completely" as Lonergan says (Lonergan, 1992, preface). Human beings are nevertheless confronted by a basic incapacity to go beyond, to understand completely, and so must try alternate procedures of self-transcendence in the face of this incapacity. In the end, as in the process of scientific judgment that relies on probability and imagination, we must extrapolate from the inner experience of transcendence to a transcendent Being who we call God.

This philosophical treatment of God through transcendence was later complemented in Lonergan's other major work, *Method in Theology*, where he developed the realms and stages of meaning in relation to feelings, experience, art, and culture. These factors came to form a more central dimension of Lonergan's argument for the existence of God. Where it borrows from science, according to Lonergan's thinking, is in theology's need for a method as successful as the scientific method. Lonergan's contribution to theology and religion is therefore based largely on his incorporation of a theory of method, which stems from his appreciation of the scientific method. For Lonergan developed a theory of how a theologian (and anyone else engaged in the humanities or social sciences) does what he or she does, a theory about the primacy of understanding the function of what it means to be a theologian. This has the effect of raising parallels between theology and science because of the similar methods used by each discipline in an engagement with particular problems in light of the history of their respective discipline.

STANLEY JAKI

One of the major writers in the science–religion dialogue, Stanley Jaki is well known for his aggressive style of argumentation. Jaki makes strong claims for the distinctively Christian origins of science. One of Jaki's central arguments is the idea that the rise of science in the West can be attributed primarily to the belief that the universe is rational. This belief was born in ancient Greece, but this was a "stillbirth." It is only through the Christian church, especially the medieval church that science fully exploded onto the scene of civilization. This amounts to a historical argument with a theological or apologetic aim. Jaki wants to convince his readers that only in Catholic Christian theology does the universe truly reveal itself as a created reality. Science, in short, is shaped by metaphysical belief in the order of the cosmos, ultimately made known in the Incarnation, God's *Logos* or Word. Specifically, this metaphysical belief opposes religious or philosophical views that stress eternally recurring cycles, such as those which held sway in China.

Citing the massive study on science in China by Joseph Needham entitled *Science and Civilization in China*, Jaki notes that Chinese thought was not hostile to science per se. But science ran into a "blind alley" in China due to "the early vanishing among the Chinese of a belief in a rational Lawgiver or Creator of the world" (Jaki, 1978:14). In contrast, the Christian west held to "a very different theological tenet, which implied the linear process from an absolute beginning, or the creation of all, to an absolute end, was the broadly shared view when science at long last found its road to unlimited advance" (Jaki, 1978:18)

Jaki is an ardent defender of the Catholic Church. In it, he sees the fulfillment of the promises made by Christ to his disciples for a tangible presence of God in the world. He sees the Catholic Church as the true defender of the Christian faith because of its dedication to truth, among other things. But Jaki's notion of truth is complex and does not align easily with the traditional Thomist school of thought that was discussed in the last chapter. Instead, Jaki takes up the thought of a number of thinkers whose appraisals of science originate in distinctly nonscientific contexts. Jaki has written, for example, on the writings of G.K. Chesterton and Henry Cardinal Newman, the founder of the Oxford movement and English convert to Catholicism in nineteenth-century England. Citing Newman, Jaki says he was a prophet for suggesting that pantheism is "the great deceit which awaits the age to come" (Jaki, 1991:13).

One scientist stands out in Jaki's mind as someone who correctly identified the relationship between Catholicism and science. This is Pierre Duhem, who was discussed in chapter 2. For Jaki, Duhem is essentially correct in two crucial ways. The first point is philosophical. Metaphysics neither constrains scientific research nor is metaphysics implied by science. This is claimed alongside a second point: the roots of modern science lie primarily in the medieval scientific concepts of impetus and inertia, not in the scientific revolution of the seventeenth century. These two medieval concepts are a necessary stepping-stone for Newton's mechanics according to Duhem. For Jaki, this is further proof that medieval science, with its ecclesial, Catholic cultural roots, is not the backward and irrational enterprise that so many modern historians deemed it. Rather, science is rooted in the scientific and philosophical accomplishments of that very ecclesiastically centered age. The Church was the original friend of science, not its enemy. The Church is the locus of belief, the expression of faith in response to revelation. As such, Jaki insists on a Catholic way of formulating the dialogue between science and Christianity:

Instead of the relation of the Bible and Science, one should ... speak of the relation of the Creed and Science. In doing so, one would also do justice to the historical reality of the former relation. Whatever concern some Church Fathers had for science, it was the Creed ultimately that they wanted to vindicate. (Barr, 1997:46)

Obviously, there is much common sense in Jaki's approach. Comparing science with biblical texts is frequently a futile exercise, given to oversimplification and mistreatment of either the biblical text or science. The (Nicene) Creed was exactly the step taken by the Church to move from a narrative account of faith toward a philosophical one. As such, it is the type of theological discourse that better lends itself to comparisons with science.

THOMAS BERRY, EDWARD OAKES, ELIZABETH JOHNSON, AND JOSEPH BRACKEN

Catholic priest and self-professed "geologian" Thomas Berry has gained a significant following amongst those who believe that the ecological crisis is a religious crisis. For Berry, as for many Catholics, one should not think about nature from a theological perspective without asking the moral question first. The moral question pertains to how human beings relate to nature, how we tend to conceptualize nature in terms of its utility for us. How do we attend or neglect our natural environment, for ourselves as individuals and as a civilization? The alleged rupture between humans and nature is a moral crisis which contains spiritual origins for Berry, who radicalizes both Catholic and ecological thought beyond what he sees as an "anthropocentric" frame of thinking. Anthropocentric thinking results when we place the needs and priorities of the human species above those of the "earth community" as he terms it.

Berry demands nothing less than a critical re-evaluation of the course of human history, Western history in particular. He adapts the genre of the philosophical interpretation of history common in Catholic thought, such as that of Augustine's belief in a six-age sequence from creation to the millennium, outlined in *The City of God*. Berry thinks western civilization's moral and ecological crisis can be judged according to the following summary:

The entire course of western civilization is seen as vitiated by patriarchy, the aggressive, plundering, male domination of our society... If we inquire into the driving forces that have evoked this critical re-evaluation of Western civilization, we can identify them as the rising consciousness of women and the devastation of all the basic living forms of earth presently taking place in consequence of the male-dominated regimes that have existed during this period. (Berry, 1988:140)

The connection between Berry's thinking and the distinctiveness of Catholic tradition is much more tentative than some of the previous thinkers that we have examined in this chapter. This has to do with the radical extent to which he foresees the need to alter our moral compass away from less important to more important issues having to do with human-nature relations.

More directly connected with the Catholic tradition is Berry's belief in the sacred character of nature. While the Catholic Church does not advocate nature's sacred character, since Christianity is not a "nature religion" per se, Berry draws on the sacramental heritage of the Church in order to claim nature as sacred. The natural world is, for Berry, "the primary revelation of the divine" (Berry, 1988:105).

Other Catholics who are concerned about environmental destruction hold similar radical views of economy and human society. They also think that the human relationship with nature cannot be altered through any shift in ideology, one way or the other. For instance, American priest Charles Murphy, author of *At Home on Earth: Foundations for a Catholic Ethic of the Environment* believes that the clash between the vices of greed and selfishness and the virtues of humility and charity are timeless and permanent. The struggle with sin and the consequent need for divine redemption suggest a more traditional orientation than a change in worldview of the sort Berry advocates. According to Murphy and Catholic doctrine, the sinful character of human existence suggests that the Catholic Church is correct in maintaining an emphasis on the role of sin and the forgiveness offered by God through Christ's saving crucifixion.

Edward Oakes is an American Jesuit whose writing on science and religion has been growing in influence over the past decade. Like other Catholic philosophers we have already examined, Oakes sees greater promise for certain philosophical categories in sorting through the debates in science and religion exchanges. For example, in a short article on the idea of the teleology or purpose of nature, Oakes adopts the scientific perspective in order to express the limits of naturalism in its attempts to state that the universe does not need a God to account for its existence. Natural scientists who try to account for the universe as purposeless are doomed to fail, says Oakes. For example, Oakes identifies Stephen Jay Gould's inability to rid the problem of how complex creatures such as human beings emerge in evolutionary history, in spite of the enormous chance occurrences that Gould claims rules out any view that humans are inevitable. Another famous scientist, Stephen Hawking, also raises the issue of teleology in nature, and he too is unable to dismiss out of hand the idea of purpose in the universe, even though he is not a theist (Oakes, 1992:540).

One reason Oakes raises the issue of teleology is due to a problem he sees when the dialogue between science and religion presupposes that science deals with "how" questions and theology asks "why" questions. Construed this way, the dialogue becomes a false dialogue between causal explanations (science) and meaning explanations (religion). This way of approaching the dialogue, as we said earlier with reference to the Thomist metaphysical perspectives, begins from the assumption that causality and meaning are not associated. But, Catholic approaches are generally much more disposed to treating life and the universe as a unified whole. Thus, Oakes claims that when scientists take teleology in nature to be a serious inquiry, they identify an end in view that is "the precursor of meaning, indeed is its very presence" (Oakes, 1992:541). What does this mean? It means that the human quest for meaning is not a reality set apart from the reality of nature. Frequently, as Oakes says, modern and postmodern philosophers separate the realms of fact and value into two entirely different arenas. The negative consequence is that the meaning of human living is taken to be an arbitrary construct that is not fully anticipated in God's creative actions. The complexity and patterns that are observed in nature as a whole are directed to the same final cause (note the Aristotelian philosophical term) as the human search for meaning itself. Oakes expands on this idea of a seamless web of reality and meaning by drawing an analogy to God's close yet noninterventionist involvement with the world as follows:

Think of primary causality, in other words, less like the ignition of a motor and more like a singer singing a song: the song is sustained only while the singer sings. But that does nothing to abrogate the laws of sound waves, of musical harmony, of the biology of vocal chords and so on. (Oakes, 2005)

So, the typically Catholic fondness for the primary and secondary causality distinction is developed by Oakes for the purpose of uniting meaning and nature, and avoiding the interventionist ideas of God and creation that are promoted by advocates of Intelligent Design.

The last issue with which we will deal in this chapter is the debate between Thomist and process thought, each of which draws a following in Catholic scholarly circles. This intellectual debate reflects the diverse concerns and tendencies that we have seen in the course of studying various contemporary Catholic thinkers throughout this chapter. On the Thomist side are those like Oakes, Lonergan, Stoeger, and Jaki whose thinking emphasizes the continuing relevance of the thought of Thomas Aquinas. On the process side are those like John Haught and Thomas Berry, whose thinking is based in part on the innovative insights of A.N. Whitehead. While there are other theological positions that have been staked out, this is the most poignant theological division in the Catholic intellectual tradition today. The debate between these two principal schools of thought in the science–theology dialogue was reflected in a discussion between Elizabeth Johnson and Joseph Bracken, SJ at a meeting of the Catholic

Theological Society of America in the mid-1990s. That debate is reflected in articles written for the journal *Theological Studies*.

Johnson, who takes a moderate Thomist position, believes the idea of creaturely participation in the life of God is sufficient to account for the fact that so many chance events occur in the natural universe. For Aquinas, according to Johnson, "nearness to God and genuine creaturely autonomy grow in direct rather than inverse proportion" (Johnson, 1996:12). Reflecting the importance of the Thomist distinction of primary and secondary causality, Johnson says that God does not work "apart from secondary causes, or beside them, or in addition to them, or even in competition with them...God makes the world, in other words, in the process of things acting as themselves" (ibid.). God is a more powerful God for doing so according to Johnson, a conclusion that is at odds with the commonsense perception of divine power: "...if God did everything directly so that created causes did not really affect anything, this would be a less powerful God. For it shows more power to give others a causative capability than to do everything oneself" (ibid.:14). She concludes her article by saying that "the basic difference between process theology and Thomism regarding God's self-limitation of omnipotence is that for process thought this is a metaphysical necessity while for Thomism it is a free and voluntary act of love" (ibid.:17).

Johnson's charge that process theology makes a metaphysical *necessity* out of God's surrender of power (through Jesus' crucifixion) is serious. She is saying that God is not free in the process theology schema. And, very few process theologians have been able to answer this charge persuasively. Jesuit theologian Joseph Bracken has done probably the best job of defending the process view of God in light of such critiques (Bracken, 1996). He does so by drawing attention to the lack of subjectivity in the Thomist tradition's interpretation of God and to the important distinction in Christian thought between God's nature and person. With the distinction between nature and person, in Bracken's view, there would be a more adequate view of creaturely freedom. Otherwise, if we stay with the Thomist view of two causal forms, that of God and creatures, we end up with the following problem: "two agents can each wholly produce the same effect only if one of them is strictly instrumental to the purpose of the other" (Bracken, 1996:723). As far as Bracken is concerned, if Thomists were not so keen on conceiving of God in objective terms, then God would not be conceived exclusively as an actual fact existing apart from the creative process of the world while (confusingly) underlying all the causal activity in the world. For Bracken, God should be understood more in terms of subjective potentiality, as one (albeit, all-important) entity in the whole "act of Being"— the universe. Bracken's portrait of God is still greatly influenced by A.N. Whitehead's elevation of the idea of creativity as something God abides by rather than being a characteristic of the universe that God authors from beyond the universe in eternity. Bracken does admit that his concept of God means that God does not know in advance what creatures will do in the future. So also is God subject to change. This is due in part to the view of chance in the universe as process theology interprets it. Needless to say, this debate raises difficult and abstract ideas about what it means to affirm that God is the Creator. This debate raises deeply abstract theological questions about what it means to affirm creaturely freedom and how God uses creaturely freedom to continuously create. It also suggests that much is at stake over how contemporary Catholic thinkers interpret and apply the historic Catholic tradition in the light of contemporary science.

CONCLUSION

As we have seen from this chapter as a whole, the variety and at times, the theological depth of contemporary Catholic thought on science and religion is broad and deeply engaging. A flurry of commentary on the Catholics and science, instigated by a *New York Times* opinion commentary by Cardinal Schönborn of Vienna, Austria, on July 7, 2005, demonstrates that a flurry of questions and doubts can arise at the slightest suggestion. In that opinion article, Cardinal Schönborn stated: "Evolution in the sense of common ancestry might be true, but evolution in the neo-Darwinian sense—an unguided, unplanned process of random variation and natural selection—is not" (Schönborn, 2005, A23). But what does "unplanned" mean? What can be affirmed as planned at the cosmic level, for example, is left unplanned at another level. The Cardinal's article and some of the ensuing commentaries on it have not necessarily kept such distinctions in mind. It may be difficult for the Catholic Church to maintain agreement on the theological implications of evolution so long as such disagreements persist about the nature of philosophical concepts like randomness. The election of Pope Benedict XVI has confirmed that the Vatican will be monitoring these discussions closely. When Benedict XVI was Cardinal Ratzinger, he took a great interest in science and his interest has continued through his invitations to several professors and former students to present papers in his presence at summer theological symposia.

It is likely, however, that the issue of interpreting science in its general theoretical setting will pale beside more pressing issues that carry moral significance. It is to such issues that we will turn in the last chapter. Ethical issues arising from the practice of genetic engineering, the question of the human soul in connection with the advances of neuroscience, and issues of the beginning and end of life are carefully studied by the Church. With some controversy, the Catholic Church has taken a variety of stances in order to draw attention to certain themes and principles in its moral teachings. Certain significant historical developments in the Church's teachings make analysis very important. Sorting out whether the Church's use of principles will withstand historical analyses and technological innovation is a key way to look at the relations between the Church and science.

NOTES

1. See EWTN website footnote to papal text at: http://www.ewtn.com/library/PAPALDOC/JP961022.HTM (accessed May 14, 2006).

2. See http://www.jesuitsinscience.org/Newsletter97/papal.htm (accessed May 14, 2006)

Chapter 6

Catholicism, Neuroscience, and Genetics

INTRODUCTION: SCIENCE AND ETHICS

The Second Vatican Council, or Vatican II as it is commonly called, ushered in a new style of theological understanding in the Catholic Church. Even though there were no major doctrinal changes made, significant practical changes have affected the way doctrines and beliefs are interpreted as truthful by Catholics. Over the course of the years 1962–1965 and presided over by two popes, the Catholic Church struck a distinct tone in its proclamation of the Christian gospel, although it is true that the church did not overturn any particular core church doctrine. Vatican II announced new ways of existing as "the Body of Christ." Called by John XXIII in 1962, the Council inaugurated a more open approach to the world. After his death in 1963, Paul VI (former Cardinal Montini of Milan, Italy) was named the new Pope. Most of the debates and discussion by the world's Catholic bishops carried on intermittently over these years focused on issues that were more strictly theological, not ethical, scientific, or even political. For example, in a document known as *Lumen Gentium* ("Light to the Nations"), issued in November, 1964, and subtitled "the Dogmatic Constitution on the Church," the Catholic bishops declared a new understanding of the Church through the concept of "the People of God," which came to replace the historically common description of the Church as a "perfect society." The Church identified itself more explicitly as a mysterious reality, an "earthly church and the church endowed with heavenly riches" forming "one complex reality comprising a human and a divine element" (I, 8 in Flannery, 1996:9).

From this significant change in expression, many expected the Catholic Church to make changes in the area of moral theology and ethics that would, in their minds, parallel the changes in ecclesiological self-identity stemming from a more open posture toward the modern world. Certainly, developments in moral doctrine are evident in a number of areas historically. The example of usury or money lending is prominent in this regard. Prior to the emergence in Europe of banking systems, the Church condemned the practice of charging interest on loans. It regarded as unjust any formal system in which unfair advantage could be provided a lending agency at the material expense of a borrower. With the advent of modern financial systems in which regulation and competition for loan making developed, the Church dropped its opposition to this practice, something modern Catholics with mortgages on their homes routinely take for granted. Similarly, many Catholics expected the Church to change its position regarding moral issues such as birth control. But this particular change would not be forthcoming, and the question that many people still ask is "why?"

Birth Control

Surely, the most noteworthy aspects of the Church's relationship with science since Vatican II have been moral and ethical issues, specifically over the use of certain forms of technology. From 1968 onward, artificial contraception and the question of birth control have remained constant in their demonstration of how the Church understands technology. Increasingly, environmental issues, abortion and then genetic therapies have taken center stage. The Catholic interpretation of birth control has embraced natural alternatives to artificial means for delaying or preventing pregnancy, by opposing the birth control pill and condoms (prophylactics). Demonstrating the gravity of the Church's view of contraception, in 1588, Pope Sixtus V issued a bull "Effraenatam," which imposed excommunication on those who engaged in all forms of contraception as well as on abortion. In part, this papal bull was based on the erroneous biological concept that the male semen contained all the active ingredients necessary for making a person. According to this view, the mother does not provide the genetic material that is now known to contribute exactly half of a person's biological makeup. The Church's reasons for opposing contraception have since shifted considerably. Many individual Catholics have also shifted their view of the question by coming to accept artificial methods of family planning, which is a shift that the Church as a whole has not made.

In 1967, Pope Paul VI appointed a group of Catholics, both clergy and lay people, to study the question of artificial contraception to see whether the Church should change its views on this question. The Pope wanted to find out whether the Church's historical proscription of artificial birth control methods was still morally compelled by Revelation. The commission was entitled the Papal Commission on Population, Family and Birth. Its task was to report to the Pope with their recommendations on whether the Church should continue to hold the view that using artificial contraception was inherently immoral. Famously, in April 1967, the commission reported that the Church could indeed change its mind. The commission's recommendations were not followed up however, when the following year, Paul VI issued a papal encyclical, *Humanae Vitae*, which argues that artificial contraception should not be practiced by Catholics, because the unitive and the procreative meanings of marriage are inseparable. Natural methods for spacing births could be adopted because these methods do not deliberately separate the unitive love of the married couple from their respective fertile potential.

John Noonan, a U.S. Appeal court judge, professor of law, and a lay American member of the 1967 commission has written an authoritative history of the Catholic Church's views on contraception (Noonan, 1967). To summarize: Noonan notes the positive value that historically the Church has placed on having children, in contrast to the hostility toward children on the part of dissenting groups such as the Gnostics and Manichees in the early church, for instance. Noonan shows that the absolute prohibition of artificial contraception has undergone significant doctrinal developments in the past due to changes in emphasis. This suggests, for Noonan, that church teaching could change in the future, at least in principle.

The teaching that Catholics refrain from using artificial contraception is based on the application of the *natural law*, which is interpreted to state, among other things, that sexual intercourse is a human activity with specifically reproductive ends that cannot be separated from the pursuit of the physical pleasure of sex. Reproductive ends are foreseen, according to the natural law tradition, in a cause/effect relationship between the act of sexual intercourse (coition) and conception (the joining of the sperm and egg to form the embryo). According to this classical interpretation of natural processes, sex is the kind of human good in which the cause of intercourse and the effect of reproduction are related as inherently necessary parts of a structural whole.

Based on the knowledge of more recent biology however, Catholic theologians have increasingly voiced dissenting views, especially since the 1960s. One prominent expression of this dissent is the now common acknowledgment that the relationship between coition and conception is statistical, not causal. That is, these thinkers argued, there is an essentially *probabilistic* relationship between the act of sexual intercourse and the randomness that governs whether or not a male sperm fertilizes a female egg in order for conception to occur. Even then, 50 percent of conceptions result in miscarriage, most of them unrecognized by the mother (Deane-Drummond, 2005:79). The logic of this criticism of church teaching is this: if nature is behaving in an unpredictable way whereby conception occurs and/or fails according to a randomly generated frequency instead of being a law-like effect of a particular cause, then there is no natural "law" that moral doctrine can attempt to uphold.

Defenders of the Church's teaching on this issue respond by drawing on Pope John Paul II's theology of the body that stresses the morality of embodiment rather than an essential morality arising from a metaphysical depiction of causes and effects. The distinct differences between John Paul II's conservatism and the older Thomistic metaphysical tradition, despite their agreement over the basic immorality of contraception, complexifies what many still take to be a simple conflict between a traditional Catholic old guard in the Vatican and married Catholics. What is clear from these discussions is that both sides are not disputing this or that scientific claim. Rather, the dispute concerns which philosophical claims about nature and ethics are valid. The main impacts of the contraception controversy have been the growing suspicion that the Church does not speak with authority on such issues and that this issue has deeply politicized and polarized Catholics. This has only spawned greater dissent from church teaching.

Many married Catholics both before but especially after the 1968 encyclical, ignored this teaching and began to use artificial contraception in order to and delay pregnancy space children. The widespread perception has been that such Catholics see no essential difference between the intention to avoid pregnancy artificially or naturally. The resulting judgment for these couples is that these methods are, in fact, morally equivalent. This view accepts the use of artificial birth control methods and is reflected in many individual and organized protests against church teaching. For instance, following a 1987 declaration by Pope John Paul II that church teaching was not open for revision, an opposing statement was issued by hundreds of theologians, known as *The Cologne Declaration*. A related issue for many dissenters from church teaching is that natural methods of birth regulation seem simply unreliable anyway, even if they were in some way morally preferable. Another argument, given the devastating toll that AIDS has taken in many parts of the world, is that the Catholic Church should sanction condom use in particular situations. For instance, argues Martin Rhonheimer:

[A] married man who is HIV-infected and uses the condom to protect his wife from infection is not acting to render procreation impossible, but to prevent infection. If conception is prevented, this will be an unintentional side-effect and will not therefore shape the moral meaning of the act as a contraceptive act. (Rhonheimer, 2004:10)

The strength of this argument is its appeal to subjective human intentionality. Proponents of natural family planning on the other hand say that, AIDS notwithstanding, much of the discussion over family planning methods is based on false assumptions. One important point of dispute that has come under critical scrutiny is that natural family planning methods are increasingly viewed as highly reliable. Advocates of the Billings Ovulation Method, for example, on its Web site (woomb.org) and through its teachers, claim over 99 percent reliability in avoiding pregnancy when it is learned and adopted in an informed way. It was developed by Australian Doctors John and Evelyn Billings who wanted to improve on the Rhythm method for the benefit of Catholic couples who did not want to dissent from church teaching. Their development of a method based on the discovery of correlations between forms of vaginal mucus and much higher fertility rates at the time of ovulation in a woman's menstrual cycle has allowed couples to use this method instead of oral contraceptives or condoms. This form of natural family planning contrasts with less reliable methods of avoiding conception such as the Rhythm method, which is based on a woman's bodily temperature readings that often do not correlate precisely enough with times of higher or lower fertility. Although practice of the Rhythm method is probably rare, it is routinely referred to as "Vatican roulette" by media outlets. Moreover, many do not appreciate that natural family planning and the Rhythm method are different methods altogether.

Pope John Paul II has dealt with this problem at a number of levels, including the 1987 ban on even official discussion of Vatican policy. He has also gone out of his way to link church teaching with his own "theology of the body," a hallmark of his thinking that dates from his earlier philosophical writings. This theme permeated his weekly homilies given at the Vatican in the first two years of his papacy from 1979 to 1981. He also met with and encouraged John and Evelyn Billings in their work.

Pope John Paul II's "theology of the body" suggests that inherent theological meaning should be attached to relationships that respect natural bodily existence and the distinctly gendered male and female sexual natures. For him, a respect for natural bodily existence and procreative capacity are critical elements in distinguishing between the moral status of regulating births through natural family planning on the one hand and artificial methods that frustrate the "natural" ends of the human sexual intercourse on the other hand. In short, although both methods can be used as ways to limit conceptions and births, they are not morally equivalent according to John Paul II. Sexuality cannot be separated from fertility. In fact, some environmentalists, including American Catholic environmentalist Wendell Berry, have also criticized artificial methods of birth control

In this January 28, 2004 meeting, Pope John Paul II meets Dr. John Billings, a cofounder along with his wife Dr. Evelyn Billings, of the Ovulation Method Natural Family Planning. Based in Australia, the World Organization of the Ovulation Method (Billings) has developed a curriculum for teaching the method of family planning that has recently become widely adopted in China. (Servizio Fotografico)

on similar grounds. For them, the prevalence of polluting materials (used condoms) and substances (chemicals present in discharged sewage) provides a distinct set of reasons for opposing artificial birth control methods. On the other hand, dissenters to the church teaching and its arguments point out that individual conscience is supremely important in determining whether and in what form birth control will be chosen. This debate has had an enormous impact on the entire generation of Catholics who came of age during Vatican II in the 1960s. It is probably true that many women in particular were alienated by what they felt was an insensitive insistence on the value of children to the detriment of the lives of parents besieged by the financial and personal responsibility that children bring. Practical considerations such as these have been mightily influential in the dissent expressed toward the Church over birth control, and they have offset the Catholic Church's increasing attachment to the value of *nature*.

The distinction between science and nature is a distinction between *practice* and *reality*. The meaning of this distinction bears on what Catholic thinkers have continued to stress in relation to a philosophy of nature. The Church's philosophy of nature, its belief in the "natural law," does have a significant, indirect bearing on science. The Church's endorsement of a largely Thomistic philosophy of nature can be seen in the Church's approach to new reproductive technologies, such as embryonic stem cell research.

First, what is the Church's teaching on natural law? The natural law tradition is rooted in the wisdom literature of the Bible, including an allusion to the truth perceived in God's creation that the wicked ignore (*Romans* 1:19–20). From the ancient stoic philosophers and the third-century Roman jurist Ulpian, the Catholic Church's understanding of natural law has emphasized two orders, that of nature and reason. These orders are a set of indicators, one physical and the other rational/spiritual in which God's creation is judged to exist through certain biological structures whose proper functioning human beings are obliged to understand correctly in order to adopt a correct moral behaviour. Traditionally, the Catholic Church has framed moral judgments in accordance with four natural law premises. First, human nature has been conceived as a universal essence, which is evident in a set of human capacities and goals. Second, the Church has understood itself as the true bearer and transmitter of moral doctrine in which no explicit interpretation of natural law was being made. Third, the Catholic position on moral questions has often evaded the question of the degree to which the Church's moral judgments were based in scripture. Protestant critics have questioned the lack of biblical references, such as the norm of Jesus' self-sacrifice, which could inform a Catholic approach to morality. Fourth, the Catholic Church has left aside many social, economic, and political factors that bear on the essential character of many moral questions by focusing on what are inherently good or evil characteristics of certain acts. According to traditional teaching, the means to an end are important in the construction of moral choice. The effects of our intentions are unpredictable, the results of speculation or moral calculus. Our own intentions, on the other hand, deal with the fulfillment and making of persons in their particular existential contexts, not definite effects from particular causes. From this brief synopsis of the Catholic moral tradition, we can better appreciate Catholic approaches to the technological advances of recent decades.

Ecology

One of the emerging areas in which natural law thought and ethics is being applied is in the area of environmental ethics. There is now a large body of literature that deals with the relationships between ecology and the world religions, especially in relation to Christianity. Since Christianity has been the most prevalent religion in the West, where industrialization took place first and most pervasively, it has borne the brunt of criticism for being instrumental in causing environmental crises. The most forceful criticism made of the Christian churches in general is that of the historian Lynn White in an article entitled "The Historical Roots of the Ecologic Crisis," which appeared in 1967 in *Science* magazine. In this article, White cites a cause and effect link between the biblical injunction to "subdue the earth" in Genesis 1:28 and the plundering and exploitation of the natural environment. The link between the biblical idea of human "dominion" over nature and environmental stress has been subsequently expressed and analyzed in many ways.

One of the enduring criticisms that has taken root in the West is the idea that the Catholic Church and other Christian churches are too focused on human concerns and ethics to the exclusion of environmental or ecological concerns. At a symposium dedicated to the science and ethics of climate change in 2007, Cardinal Martino of the Pontifical Commission for Peace and Justice was quoted as saying that "Nature is for man, and man is for God." This is further evidence of the anthropomorphic interpretation of the relationship between nature and human beings that has been a theological mainstay throughout church history.

However, there is a widespread view now that the metaphorical language of the Bible, which depicts creation in anthropomorphic terms has cast Christianity as too "human" a religion. For instance, in Romans 8:22, creation is said to be "groaning in labour pains" along with the children of God in anxious expectation of a redemptive response by God in history. The critique of the Church over its alleged excessive focus on human affairs has led to a growing ecological theology movement. In fact, even before Lynn White's important essay, theologian Joseph Sittler spoke at the 1961 Third Assembly of World Council of Churches and called pollution a form of blasphemy (Scharper, 1999: 226). Having flourished into a diverse movement, many are persuaded to think of ecological theology and its twin, ecological spirituality, as akin to a new religion. Nevertheless, the presence of many devout Christians, Jews, and Buddhists within the ranks of environmental movements renders such assessments incomplete, at best.

The most forceful environmental messages from within the Catholic Church have come from two sources: papal teachings and bishops' statements. In the category of papal writing, John Paul II's encyclical, *Centesimus Annus*, written in 1991, is of prime importance. It was written to mark the centenary of Catholic social teaching, first developed explicitly by Pope Leo XIII in 1891 in the encyclical *Rerum Novarum*. He notes:

Equally worrying is *the ecological question* which accompanies the problem of consumerism and which is closely connected to it. In his desire to have and to enjoy rather than to be and to grow, man consumes the resources of the earth and his own life in an excessive and disordered way. At the root of the senseless destruction of the natural environment lies an anthropological error... Man thinks that he can make arbitrary use of the earth, subjecting it without restraint to his will, as though it did not have its own requisites and a prior God-given purpose, which man can indeed develop but must not betray. Instead of carrying out his role as a co-operator with God in the work of creation, man sets himself up in place of God and thus ends up provoking a rebellion on the part of nature, which is more tyrannized than governed by him. (*Centesimus Annus*, #37)

In this passage, John Paul II states more explicitly the moral terms of the environmental crisis. Moreover, the root problem for John Paul II is not the anthropocentrism of the Genesis creation narrative but rather the human failure to adequately understand it. Church teaching therefore reiterates the undeniable role of moral deliberation in reckoning with the environmental crisis. Moral deliberation has greater significance than worldview, the question of whether one's view of the world is too focused on human affairs. For John Paul II, the environmental crisis, while broader and alarming in its sheer scope, is nevertheless very much like other human social crises in its being rooted in human sin, especially greed. This encyclical followed five years after a historic interfaith prayer gathering held in Assisi, Italy, where the Pope was joined by the Archbishop of Canterbury, the Dalai Lama, and other world religious leaders. Assisi was chosen in part for being the locale where Saint Francis (dubbed the "patron saint of Ecology") cultivated a following for his spirituality of animals and nature in the early thirteenth century. Aware that the concern for nature has moved to the center of global attention, John Paul II reflected regularly on the role of nature in the Christian view of God: "Nature therefore becomes a Gospel that speaks to us of God: 'For from the greatness and beauty of created things comes a corresponding perception of their Creator' (Wis 13:5)" (John Paul II, Jan. 26, 2000).

Beginning in the 1980s especially, Catholic expressions of concern over environmental crises have become commonplace at the level of bishops' conferences, which are groups of bishops working together on regional, national, or continental assemblies. Examples of such concern have come from the Filipino bishops in their 1988 statement titled "What is Happening to Our Beautiful Land?" and similar statements by bishops in the Columbia River watershed in the U.S. northwest and the Canadian Conference of Catholic bishops in regard to development in ecologically sensitive areas of northern Canada.

BIOETHICS, PERSONHOOD, AND LIFE

Moral questions such as birth control and the ethics of war have dominated post-Vatican II church discussions with science and social ethicists. But increasingly consequential is the growing attention paid to the ethics of the human person. Controversies surround the physical beginning and end of human life. The status of the embryo in the abortion debate is the most prominent issue in the field that is termed "bioethics." The term bioethics was coined by the Dutch thinker van Rensellar Potter, who conceived of it as something that would encompass an ethic of the land, ecology, and human consumption. However, as the term came into wider use, the field became focused on much narrower issues that pertain to human life directly. More specific still is the perennial issue of the status of the human embryo with which much of bioethics is recently preoccupied. But, this issue extends beyond abortion since that the human embryo is also implicated now in stem cell research. A related perennial question that has received growing attention in recent years is euthanasia. It too involves the question of whether and why human life can be ended intentionally with moral justification. The latter issue tends to involve assessments of personhood, yet without the same level of scientific debate that shapes discussions surrounding stem cell research and abortion. In order to provide assistance to the Church in negotiating the thicket of complex ethical issues, Pope John Paul II issued a *motu proprio* on February 11, 1994, instituting the Pontifical Academy for Life. Its objectives are "the study, information and formation on the principal problems of biomedicine and of law, relative to the promotion and defense of life" (Pontifical Academy of Life).

At the heart of the Catholic Church's position concerning the issue of human life is the question of human dignity and the soul. The Catholic Church claims, in contrast to certain other secular frameworks for conducting bioethical analyses, that human life is not something one achieves on one's own but is rather something one receives as a gift from God. This in itself does not necessitate any one moral position concerning abortion or euthanasia, but it certainly sets in motion different sets of questions about what may or may not be morally permissible. Absent this starting point about the nature of human life, moral inquiry takes on a wholly different character. Catholic positions on abortion now take it for granted that the Church teaches that human life begins at conception. But, ignorant of modern embryology, the Church historically was more ambivalent than it is now. One popular view during the Middle Ages, for example, was that human life begins at the moment of "quickening," the time several weeks into a pregnancy when the mother first feels the movement of the child. The widely held view before the Protestant Reformation is that at about 40 days after conception, the human soul is created by God directly and infused with a human embryo to create a person. Aquinas held a complex version of this view with a succession of vegetative, animal, and then rational souls directing the emergence of a fully human person. Scholars investigating Aquinas' view debate whether, for example, he was faithful to Aristotle's view of when a human person comes into being. Even though that historical debate deals with what Aquinas thought, it is essentially duplicated in debates amongst Catholic ethicists today about whether life begins at conception or around the fourteenth day after conception, at which time it becomes impossible for an embryo to twin and become more than one individual. Those who believe that the human person begins around the fourteenth day do so because the possibility of twinning prior to that point prevents one from identifying the early human embryo as one *individual* human being.

Stem Cell Research and Cloning

One of the most prominent bioethical issues to arise since the late 1990s is that of stem cell research. Stem cells are undifferentiated precursor cells that can be either perpetuated as more stem cells or reproductively differentiated into specialized types of cells. The Catholic Church, its teaching bodies and bishops, have criticized stem cell research that involves the destruction of human embryos. On the other hand, it has welcomed and encouraged stem cell research that does not destroy human embryos. Since the 1990s when technology advanced to a point where stem cells could be harvested from embryos, human embryos were thought to be the most valuable source for stem cells because of their radical multipotency or multiple uses. However, since the ban on embryonic stem cell research announced in the United States by President George Bush in the summer of 2001, more research has gone into the development of therapies and techniques for manipulation of adult cells for the creation of stem cells, a procedure that does not eliminate human embryos.

Some Catholics have gone so far as to actually fund such research. According to one Reuters report (March 24, 2005) the Catholic Archdiocese of Sydney, Australia, supported researcher Alan Mackay-Sim of Griffith University in the creation of stem cells taken from a human nose. According to Mackay-Sim, "the adult stem cells taken from inside the nose could potentially be used to grow nerve, heart, liver, kidney and muscle cells." Another proposal, still in its early conceptual stage at this point, is to create multicelled organisms with embryo-like properties but that are not actual embryos. Embryonic stem cells could be harvested for the development of therapies for certain degenerative diseases. One of the most prominent proponents of this research is William Hurlbut, a member of

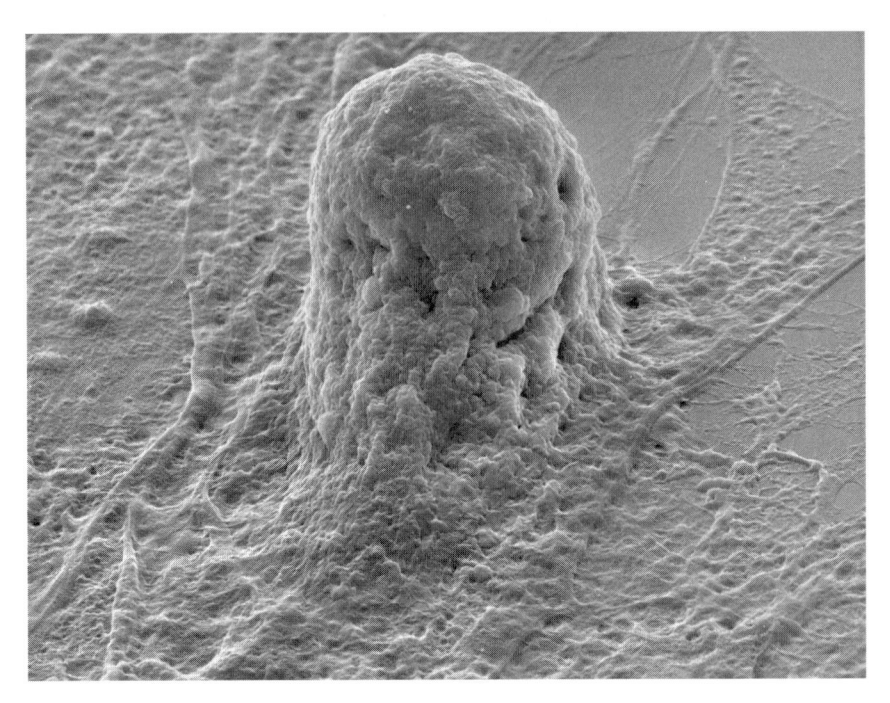

Human embryonic stem cell (gold) growing on a layer of supporting cells (fibroblasts). Stem cells are derived from very early embryos and can be either grown to stay in their original state or triggered to form almost any type of human cell. The fibroblasts provide special factors that maintain the stem cells in their original state. The stem cell appears to be grasped by the underlying fibroblast. Stem cell research could lead to cures for many diseases such as Parkinson's disease, Alzheimer's disease and diabetes, where cells are damaged or absent. Color-enhanced scanning electron micrograph by Annie Cavanagh and Dave McCarthy. Stem cells were grown by Jessica Cooke in Stephen Minger's labs at King's College London. (Courtesy of the Wellcome Trust)

the U.S. Presidential Commission on Bioethics, who said in 2004 that "it may be possible to produce embryonic stem cells within a limited cellular system that is biologically and morally akin to a complex tissue culture and thereby bypass moral concerns about the creation and destruction of human embryos" (PCBE, December 3, 2004). Another research success along the same lines was publicized in 2006 and announced by Robert Lanza of Advanced Cell Technology regarding a procedure in which stem cells are extracted when the early-stage embryo only consists of between 8 and 10 cells. On the day that the research was publicized, Lanza was quoted as saying: "Embryonic stem cells have been synonymous with destroying life... But if you're not hurting the embryo, why wouldn't you be able to move ahead? We think this will solve the current political impasse" (Toronto Globe and Mail, August 24, 2006).

Catholic responses have noted that avoiding destruction of embryos for the purpose of stem cell harvesting is based on a crucial ethical distinction. The operative distinction is between using materials from part of a thing that is not destroyed in the process of harvesting on the one hand and destroying a living thing on the other hand. In the case of the human embryo, the living thing is claimed to be a human organism and so it has what is termed "inherent" or "intrinsic" value. The embryo, if left in the womb and nourished there in its natural environment, will set its own equilibrium or what biologists call "homeostasis." On the other hand, parts of organisms or information about organisms that do not have that self-organizing capacity cannot become anything without being deployed through human artifice within and for another organism. According to natural law tradition that has been the standard framework for Catholic thinking about moral questions, an embryo has the dignity accorded to human beings because of this self-organizing potential, its "thingness." It naturally obtains matter from the external environment, the mother's body, in order to become what it already is: a human being.

Some debates occur between persons who accept the basic philosophical distinctions but who disagree about when personal identity takes place, with some arguing that individuality is present before 14 days after conception with others arguing that individuality can only be established after the 14-day mark. As discussed already, the main reason behind the argument for individuality occurring after 14 days is the occurrence of identical twins, whose individuality is clearly not established before this date. Overall, what is striking about the debate over stem cells and the proposals designed to avoid the destruction of stem cells, is the reliance on Aristotelian language of "things" and "potentials" by those who advocate preserving life. This tendency on the part of some bioethicists reflects concerns shared by the Catholic Church for grounding in an account of nature the idea of human dignity and the equality of all persons. It forms the key plank of the natural law tradition that claims that values and ethics need to be grounded in nature as opposed to being created, free-floating sets of ideas beyond nature. Natural law, not Catholic doctrine bases Catholic arguments against human embryo destruction in stem cell research or destruction of the fetus in abortion.

The cloning of human beings is another technologically aided breakthrough technology about which the Catholic Church has serious concerns and made several interventions. Beginning with the birth of Dolly the sheep in February 1997, the first successfully cloned mammal, scientists and agencies worldwide have worked to bring about an international consensus against the possibility that the cloning of human beings might come about. Concern over the condition of cloned creatures, whether human or not, mounted in 2003 when Dolly died prematurely of arthritis. The speculation is that cloning would lead to much greater disease, suffering, and premature death than for organisms created under natural circumstances (O'Leary, 2006:240).

Cloning is related to embryonic stem cell research because one method of extracting embryonic stem cells is from embryos cloned for that purpose alone. This is a process that is specified as "therapeutic" cloning. Conceptually, this is different from reproductive cloning, the express purpose of which is to reproduce an identical organism. Writing about reproductive cloning, the Pontifical Academy of Life issued a statement in 1997. While cloning cannot create a soul or the unique personality that arises during a person's human development, this document links cloning to: eugenics, the problem of confused parentage and the attempt to subvert God in Western civilization. Moreover, the academy invokes two principles dear to secular political philosophy in arguing against cloning: the principle of the equality of human beings and the principle of nondiscrimination, both of which could allegedly be undermined by the selective aspect of cloning. On this view, cloning is thought to lead inextricably to the domination of some human beings over others, those engaged in cloning and the cloned. (O'Leary, 2006:234).

Euthanasia

Euthanasia has become controversial in recent years for related reasons. The liberalization of laws concerning euthanasia in the state of Oregon in the United States, Belgium, and the Netherlands has drawn much attention. Less well known but important in Catholic thinking is a 2004 allocution by Pope John Paul II mandating that family members and medical caregivers are morally obligated to provide artificial nutrition and hydration (ANH) to persons who cannot regain consciousness, who are in what are called "persistent vegetative states" (PVS) (Pope John Paul II, March 20, 2004). The case of PVS persons rocketed to worldwide prominence in 2005 with the case of Terry Schiavo in Florida. The significance of the 2004 papal statement is its insistence that providing nutrition and hydration should *not* be considered a medical procedure. This statement is based on the distinction between ordinary and extraordinary means of preserving life. Emerging within papal teachings over the course of the 1980s and 1990s (in such encyclicals as *Evangelium Vitae*—"The Gospel of Life") is the idea that "if death is not imminent then life must be prolonged" (Sowle Cahill, 2006:129), a position that liberal critics call the "idolization" of biological life. The 2004 statement is a culmination of this trend that is related to John Paul II's valorization of human suffering, the idea that in suffering, we may participate in the suffering that Christ endured on the cross. It identifies ANH clearly as an "ordinary" means of preserving life in order to support patients whose lives might otherwise be deliberated within a more utilitarian logic about whether or not the patient is enjoying a "quality of life" or not. The 2004 statement spurred a debate that has dealt with the relationship between biological life and the spiritual life. On the one hand are those who agree with John Paul II that life must be maintained, even if it is obvious that those who exist in persistent vegetative states cannot appreciate or act on any of the spiritual goals of life. On the other hand are Catholics such as Kevin O'Rourke and Lisa Sowle Cahill. O'Rourke believes that the capacity to pursue the spiritual ends of life is a necessary capacity for mandatory ANH provision. What this means in fact is that ANH is not mandatory once the functioning of the neocortex of the brain is irreversibly lost. In other words, there must be a better than zero chance that the ill person will recover some measure of consciousness. Sowle Cahill goes further by arguing:

An egregious insistence on ANH in virtually every case may be intended to protect vulnerable persons from utilitarian cost-benefit calculations regarding the care appropriate for them. In reality, it is likely to enshrine inflexible regimes of 'treatment' that few people would choose for themselves, that contradict best interests and humane care of the ill and dying, and that turn a blind eye to the real and immediate health needs of the many who cannot access even basic and useful care. (Sowle Cahill, 2006:132)

As Sowle Cahill notes, one of the most underdeveloped areas of discussion in bioethics is the dimension of justice, which ironically, is developed in more detail in regard to social issues in Catholic social teachings. While John Paul II's 2001 call for President George W. Bush to protect human embryos in regard to stem cell research was widely publicized, his earlier call for Bush to propel a surge in opportunity for the world's poor to improve their lives was virtually ignored.

Sociobiology

The theme of justice brings us to the theme of love, which has become the subject of scientific attention over the past few decades as evolutionary theory has developed in sophisticated ways. In the nineteenth and twentieth centuries, evolution's reputation for portraying nature as "red in tooth and claw" was taken by Herbert Spencer and practitioners of eugenics, for instance, as showing that nature is merciless and vicious. Spencer famously applied this mercilessness as the basis for promoting the human species at the expense of the weak or "bad" in society: "to aid the bad in multiplying, is, in effect, the same as maliciously providing for our descendants a multitude of enemies" (Spencer, 1874:346). This emphasis on competition has since mellowed. A complementary emphasis on cooperation has emerged amongst practicing biologists and philosophers of biology. Evolution has been increasingly interpreted to account for a wide array of social and political structures, and this what is studied in "sociobiology." Closely associated with the American entomologist E.O. Wilson, this hybrid discipline analyzes how certain behavioral patterns are rooted in evolutionary history for promoting an individual's survival and reproduction. Sociobiologists place a great deal of trust in the significance of genes and gene sharing to tell the story of evolution. This is why sociobiology is thought to be so inherently reductionistic. For instance, there is a tendency amongst sociobiologists to think of emotional responses as genetically programmed instead of being the result of a complex interplay of a large number of genetic, environmental as well as intentional factors. William Hamilton coined the term "inclusive fitness" to account for reciprocal behavior amongst relatives—who share genes by definition. This sharing of genes decreases rapidly as we move from relationships in which 50 percent of genes are shared (between brothers and sisters, for example) to those in which only 25 percent of genes are shared, as with aunts and nephews. "Kin selection" was introduced to understand the behavior of younger animals who aid their parents in raising other brothers and sisters instead of going out and starting families of their own. Ornithological studies have documented this behavior pattern. "Reciprocity" is behavior in which a donor makes a small loss for a recipient's more significant gain with the expectation of future benefit. This kind of behavior pattern has been observed among bats. And general "group selection" tendencies have also been noticed amongst species where "alarm calls", for instance, are issued to warn a larger population of a danger at the risk of an individual's own safety. Such conceptually distinct behaviors have given rise to the hypothesis that empathy—basic regard of human beings for others—can be explained in evolutionary terms.

What does this mean for the Catholic Church, which proclaims the love of God through the person of Jesus is the unique model for human love? In other words, do scientific studies of the evolutionary, instinctual nature of love contradict the Church's claim that love is primarily a supernatural reality? Theologian Stephen Pope has responded to the sociobiological research programme by sketching a portrait of love that is multilayered and differentiated. Love is an ordered, stratified hierarchy and evolution has "bequeathed human nature the capacity to transcend . . . its own basic framework" (Pope, 1994:154). However, for Pope, evolution and the natural capacity for altruism makes claims on our ethical capacities too. For him, Catholic moral thought should extend beyond a preoccupation with the ethics of friendship and social concern to take into consideration the ways in which human nature predisposes human beings to love in predetermined ways that are not fatalistic or contradictory of human freedom (Pope, 1994:10).

CATHOLICISM, EMERGENTISM, AND BRAIN SCIENCE

The importance of the human soul has already been discussed with regard to the beginnings of human life. As Catholic theology has traditionally held, the person is a creature made up of a soul and a body. The importance of the soul was signaled in the 1996 statement issued by John Paul II on evolution, the same statement in which he stated the Church's recognition that evolution was more than just a hypothesis. John Paul II stated in that address that "the human body evolved taking its origin from pre-existing living matter, [but] the spiritual soul is immediately created by God" (John Paul II, "Message to Pontifical Academy of Sciences on Evolution," 1996b:351). Anne Clifford of Duquesne University calls this the Catholic version of creationism (Clifford, 2004:295). However, understanding the meaning of the human soul has also been deeply informed by debates in both the neurosciences and the philosophy of neuroscience, as well as debates over the definition of human death.

Following the enormous strides that have been made in neuroscientific studies since the 1980s in particular, some have noted the resulting challenge for Christians in general:

[A]s the neurosciences now increasingly throw their considerable weight behind arguments for the unity of human being—a purely physical organism—they have called theology's comity into question... if God has to do with humans at all, it must be by means of interaction with their bodily substance and, more particularly, with their brains. (Murphy, 1999:iii)

Proposals for the physical unity of the embodied person insist that no immaterial substance or principle exits apart from the human body, a position clearly at odds with historic Catholic doctrine. The idea of the soul's immediate creation arises early on in the Church's adaptation of Plato's philosophy of the soul, which treated the soul as an entity distinct from the human body and as the primary agent in human action. In speaking of Socrates in his dialogue *Phaedo,* Plato speaks of the soul as "the invisible part, which goes away to a place that is, like itself, glorious, pure and invisible" (Plato, 1955:79; §80d). The influence of Platonic philosophy on Christian thought has been enormously significant in the history of the Church. This was especially true for early church fathers, such as Origen, who even postulated the idea that creation itself was the result of sin, rather than the more orthodox view held by Irenaeus that the physical world was created purposely by God. The soul's immortality is also associated with the adoption of Platonic philosophy.

How do Catholic scholars respond to the challenge from physicalism, the position that the human person *is* the human body? Essentially, there are two basic counter-proposals. First, some insist on the invisibility of the soul itself and speak of the soul as a principle, not a substance. Such a principle is what can account for human agency, a Thomist emphasis. If we were to understand the soul as a Cartesian substance entirely separate from the material body, then there are obvious dilemmas for speaking of the human person as a unified being. The notion of a principle is thus thought to be superior to that of substance amongst most Catholic thinkers. Principles fit with what is termed an emergentist portrait of naturally existing entities. Emergentism is an interpretation of structural sciences like evolutionary biology and human biology, in which core properties that determine the functioning behaviours of a system cannot be inferred from an analysis of constituent parts of the system. For example, if it is impossible to understand the behavior of water by examining the properties of hydrogen and oxygen separately, then, so the argument goes, it is impossible to understand the human person as a whole by understanding the various brain networks, lobes, neurons and so on—separately. As a principle, the soul is a concept that explains the unity of consciousness and all its powers. Human consciousness is something different than other forms of consciousness or awareness, and the soul is one way to articulate this difference while maintaining the essential unity of human cognition that physical arguments tend to stress.

The other way that the challenge of the neurosciences has been handled is through the discussion of death and exactly when death occurs. The Pontifical Academy of Sciences has dealt with this issue through studies by a working group in 1989, whose job was to try to determine the relationship between "brain death" and the death of the human person. It concluded that brain death should be regarded as the death of the human person. This issue is not mundane due to the great need to retrieve human organs for transplantations at the precise moment when it is medically advisable to do so shortly after death has occurred. Consonant with the physicalist emphasis on human embodiment, the group defined death clinically as "when there has been total and irreversible loss of all capacity for integrating and coordinating physical and mental functions of the body as a unit" (May, 2000:289). The central presupposition of this report and a near consensus of other scientists, Catholic and non-Catholic, is that the brain is the central integrating organ of the human body. Whatever concept one has of the soul, it is virtually assumed that the soul is a concept that pertains first and foremost to the brain as the central integrative organ of the human body. This assumption is consistent with an emergentist portrait of the human person in which a unified hierarchy of natural mechanisms make up the human body, with the brain acting as a central organizing mechanism at the center of the "system." The soul is a concept that designates the agency that marks the brain in operation, and in that regard the soul can be seen as an "emergent property" of the bodily system. Discussions within the Pontifical Academy are testimony to the central place now accorded the brain in understanding human life. In this particular case, the Pontifical Academy was able to clarify that artificial means taken to prolong cardiac functioning in order to preserve bodily organs for some time after death for the purpose of transplantation is morally permissible.

CATHOLICISM, GENETIC SCIENCE, AND ORIGINAL SIN

The last issue that we will deal with here is the increasingly contentious area of genetic science and the technological advances that pertain to genetics. As we have already seen with regard to stem cell research, for instance, contemporary genetics contains two broad sets of implications, one ethical and the other theological. In the case of embryonic stem cell research, there is both a question about what research may be permitted and about when human personhood begins.

Celia Deane-Drummond is one Christian ethicist who has integrated core aspects of the Catholic theological tradition with an attempt to methodically deal with the numerous ethical challenges brought on by genetics. Rejecting what she calls a rule-based approach to ethics, Deane-Drummond says the best way to handle the complexity of these issues is to have recourse to the four classic or "cardinal" virtues that may guide ethical deliberation. These virtues are: prudence, justice, fortitude, and temperance. These philosophical virtues complement and mediate the theological virtues of faith, hope, and charity that direct the Christian life. Building on the reflections on human freedom by Catholic theologian Karl Rahner, Deane-Drummond seeks to pinpoint ethical ways forward in the field of genetics while living "in the shadow of eugenics." Eugenics is the attempt, pursued by the Nazis and others, of trying to eliminate certain individuals from mainstream society by preventing them from reproducing. Francis Galton, who was Charles Darwin's cousin, has been called the father of modern eugenics and his definition of eugenics refers to the "science of improving stock" and of giving "the more suitable races or strains of blood a better chance of prevailing over the less suitable" (Deane-Drummond, 2005:56).

There are two kinds of genetic therapies under consideration in contemporary medicine. One is somatic gene therapies in which genetic alterations are carried out only on an individual. Germ-line cell therapies, on the other hand, are those that involve alteration of sperm, ova, or their precursors and that will affect not only the individual but all of his or her descendants. Genetic testing, screening, genetic counselling, and gene patenting are emerging practices that are laying the groundwork for genetic therapies and, as such, they place enormous responsibilities on the shoulders of policy makers. They are chiefly concerned about what ethical guidelines need exist to ensure how genetic "enhancements" are practiced. For instance, protocols have to be established concerning the disclosure of genetic information gathered through testing, which is particularly onerous, since possession of such information by insurance companies can result in discrimination by employers or insurance companies themselves. This becomes more complex if, for instance, a child needs to be tested. If the test is for a childhood disease, this is straightforward, but if the test is for a "late-onset" disease, Deane-Drummond suggests that it is just to disallow such tests to be conducted on children since disclosure (intentional or accidental) of the test results might result in discrimination against that child later in life. Of course, such a policy is a curb on the freedom of that child and her parents in the meantime. This is coercive on the part of the state or the justice system if such a ruling is backed up by judicial ruling and depending on whether one has a positive or negative assessment of the role of the state in society, one's tolerance for curbs on genetic testing will be affected.

The role of virtues is particularly relevant for dealing with the arrival of genetic information banks and the counseling and possible therapies for various conditions currently being researched and developed. For instance, Deane-Drummond sees prudence as a virtue that encourages stillness and silent contemplation over the "impulsive reaction to perceived threat." For families living with disabled children or devastated by genetic disease, many come to see the "face of wisdom" while others believe that "the virtue of wisdom is insufficient" so that "the virtue of charity comes into its own, for no family or individual is an island; and the task of Christian community is one of solidarity" (Deane-Drummond, 2005: 123). The themes of love and solidarity are prominent in the social teaching of the Catholic Church. Deane-Drummond's recourse to such virtues as providing approaches to issues in what is termed "genethics" can be seen as a characteristically Catholic response. Positing a virtue ethics for dealing with genetics is thus consistent to a large extent with the traditional natural law framework for understanding moral issues, even though virtues by themselves are less constrictive on the range of possible alternatives available.

The other impact of genetic science is properly theological. Since Saint Augustine, the Catholic Church has held to a concept of original sin in which Adam's personal choice to sin against God has affected all of humanity through a biological transmission of sin from generation to generation. The objective of the doctrine of original sin is to explain how it is possible that sin is, in the language of the Catechism of the Church "an offense against reason, truth, and right conscience." It goes on to agree with Augustine that sin is a deprivation that is manifest in "an utterance, a deed, or a desire contrary to eternal law" (CCC, 1849). The dilemma for the Church is that God created the world as good yet sinfulness is pervasive. Sin is so pervasive that the twentieth-century theologian Reinhold Niebuhr has called it one of the most empirically verified facts of human existence. Based on Paul's development of a contrast between Adam and Christ in Romans (chapters 5 and 8), the Christian tradition has understood redemption in Christ as healing the sin that passed down from Adam, a form of defect that even affects the rest of creation.

The Christian Church was not predisposed toward this view universally from the beginning however. For example, Theodore of Mopsuestia, a third-century church father, believed that "it is only nature which can be inherited, not sin, which is the disobedience of the free and unconstrained will" (Pelikan, 1971: 285). This disagreement over how Adam's sin is transmitted to the rest of humanity has been the source of a number of related controversies, among which is a distinction between the Eastern Orthodox and the Catholic Church on the identity of the human *imago dei*. For the Orthodox, the human person, body and soul is an icon, the image and likeness of God, not the soul alone, which is what they allege Augustine says (Ware, 1997:220). For Augustine, it seems that Adam's sin casts a darker shadow owing to his own conception of the perfection and harmony in which Adam and Eve enjoyed in their original bliss.

Perhaps the tension between East and West is best characterized in light of how one Orthodox theologian regards Adam: "[he] fell, not from a great height of knowledge and perfection, but from a state of undeveloped simplicity" (Ware, 1997:223). To some extent, the "eastern" interpretation of original sin, which also involves no personal sense of inherited guilt for what Adam did, may be regarded as more evolutionary than "static." To some extent then, this interpretation may be more conducive to science. Nevertheless, it is widely acknowledged that Augustine's thought was retrojected onto him by post-Reformation theologies that possessed a dim view of human potential while drawing inspiration from Augustine.

Original sin is meant to be a doctrine that explains Adam's (and Eve's) sin as a form of disobedience that brought humanity out from an original state of holiness into a state of failure. Some contemporary theologians such as Lutheran Philip Hefner prefer to think of original sin as a discrepancy between the information arising from our genes and the demands of our cultural and social contexts. Sinfulness, in Hefner's view, is better understood as finitude or learning by trial and error. Catholic theologian Denis Edwards, responding to the opposition between the Augustinian tradition and Hefner's rejection of sinfulness, sees both finitude and sinfulness as essential categories for understanding the human person. As he says:

> the discrepancy and fallibility described by Hefner are real, but they are not of themselves sin... Karl Rahner has offered a major clarification of the theological concept of concupiscence... a clear distinction between concupiscence as "the child of sin" and concupiscence as the "companion of finitude." [So] it is important to avoid any tendency to identify selfishness and sin with the biological side of the human being and unselfish behavior with the cultural side... Original sin [is] the impact of accumulated history of rejections of God... [it] may involve a lack of acceptance and integration of [either] the biological side [or] forms of cultural conditioning. (Edwards, 1998:384–385)

As Edwards emphasizes, it is important not to react to the overly simplistic Augustinian notion of genetic inheritance of sinfulness with equally simplistic concepts of human biology that would divorce genes from sinful behavior or simply redefine sin as the limited capacities of genes. Again, like Deane-Drummond, while Edwards is developing his own systematic and conceptual frameworks, his response to the problem of genetics in the context of original sin is both coherent with the tradition while being a revision of that framework in light of recent scientific advances. In Edwards' revision of Augustine, through the genetic inheritance of the tendency to sin, or concupiscence, there is no simple cause and effect transmission, but rather an inheritance of ranges of possible behavior patterns that can be improved upon or worsened, depending on the culturally influenced choices that mark a person's life.

CONCLUSION

In this chapter, we have examined a host of contemporary issues that are permeating ethical and even theological discussions and that push all religious traditions, including the Roman Catholic Church, to respond. Such responses, as can be seen with regard to contraception, stem cell research, and genetic science call on academics and the Church to engage in highly interdisciplinary and creative ways. It is for the observers to judge which actors in this ongoing drama are most successful in surmounting multiple, overlapping challenges.

The twentieth century has seen a slow series of interconnected movements within the Catholic Church and Catholic thought to renew and develop a catholic relationship with science. It is difficult if not impossible to convey the meaning of these interconnections within any single concept or in reference to any one historical event. At the scholarly level, much material has been, yet still needs to be, sorted out. For example, as the Schönborn episode and the contraception controversies show, it is often challenging to distinguish between the Church's perception of the way nature operates and individual Catholics' perception of nature or natural law. It is also sometimes challenging to distinguish between the use of natural categories to justify a theological, a philosophical, or a moral stance (or combinations of such stances) that could, in principle, be shared with non-Catholics. Nevertheless, given the much more concerted collaboration amongst Catholics in the science–religion dialogue and between Catholics and non-Catholics as well, we can be much more confident than forty years ago that the range and coherence of Catholic voices in this dialogue will make its mark over the first half of the twenty-first century. Given the worldwide presence of Catholics in a large number of fields, and embedded as many of these scholars are in the Church and Catholic as well as secular universities, it is much more likely that Catholics will impact the science and religion dialogue in the next few decades in a more substantial way than in the period from Vatican II until the present. The sometimes unique stances that Catholics adopt on moral and ethical issues will certainly ensure that Catholic voices in the dialogue will be more distinct and therefore more visible.

Overall, one is permitted to see the future of relations between Catholicism and the natural sciences as subject to almost innumerable forces and subject to many implications arising from rapid scientific advances. Perhaps the most serious areas of contention are the nonphysical component(s) character of the human mind and the potential for discoveries concerning parallel universes. Developments in either of these two subjects have more immediate consequences for the Catholic Church and Catholics in the sense that there are immediate consequences for church teaching. As we have seen in this book, the contrast between the Catholic Church's centralized authority and the multiple centers of authority in science accentuate what is already a wider tension between scientific and religious claims in the contemporary world. The goodwill and scholarship exercised by many Catholics, including church authorities, has gone a long way toward mitigating the effects of the Galileo controversy and the minor crises of earlier times. Whether relations between Catholicism and the sciences will continue to move in this direction remains to be seen. Worldwide cultural trends as well as internal church struggles will both have an impact on whether this happens.

Primary Sources

—1—
Galileo Galilei, Letter to Grandduchess Christina, 1615

In a dinner table exchange at the de Medici court, philosophy professor Cosimo Boscaglia argued that heliocentrism could not be true because it was contrary to the bible. Galileo's former student Benedetto Castelli (now a Benedictine monk) was called upon to answer the scriptural arguments brought against Copernicanism by Christina of Lorraine, widow of Duke Ferdinando de Medici. Galileo learned of the incident and composed an open letter to Christina that was widely circulated (although not published until 1536). Galileo's intent was to offer an exposition of the principles of biblical interpretation that he believed supported his conviction that heliocentrism was not in conflict with those few scriptural passages touching upon cosmology. Although Galileo's exegesis sounds surprisingly contemporary to us today, in 1615 he was courting danger by trespassing (as a mere mathematician) upon the professional turf of biblical theologians.

I. The Authority of Scripture

Some years ago I discovered many astronomical facts till then unknown. Their novelty and their antagonism to some physical propositions commonly received by the schools did stir up against me many who professed the vulgar philosophy, as if, forsooth, I had with my own hand placed these things in the heavens to obscure and disturb nature and science. These opponents, more affectionate to their own opinion than to truth, tried to deny and disprove my discoveries, which they might have discerned with their own eyes; and they published vain discourses, interwoven with irrelevant passages, not rightly understood, of the sacred Scriptures. From this folly they might have been saved had they remembered the advice of St. Augustine, who, dealing with celestial bodies, writes: "We ought to believe nothing unadvisedly in a doubtful point, lest in favour of our error we conceive a prejudice against that which truth hereafter may discover to be nowise contrary to the sacred books."

Time has proved every one of my statements, and proving them has also proved that my opponents were of two kinds. Those who had doubted simply because the discoveries were new and strange have been gradually converted, while those whose incredulity was based on personal ill-will to me have shut their eyes to the facts and have endeavoured to asperse my moral character and to ruin me.

Knowing that I have confuted the Ptolemaic and Aristotelian arguments, and distrusting their defence in the field of philosophy, they have tried to shield their fallacies under the mantle of a feigned religion and of scriptural authority, and have endeavoured to spread the opinion that my propositions are contrary to the Scriptures, and therefore heretical. To this end they have found accomplices in the pulpits, and have scattered rumours that my theory of the world-system would ere long be condemned by supreme authority.

Further, they have endeavoured to make the theory peculiar to myself, ignoring the fact that the author, or rather restorer, of the doctrine was Nicholas Copernicus, a Catholic, and a much-esteemed priest, who was summoned to Rome to correct the ecclesiastic calendar, and in the course of his inquiries reached this view of the universe.

The calendar has since been regulated by his doctrine, and on his principles the motions of the planets have been calculated. Having reduced his doctrine to six books, he published them under the title of *De Revolutionibus Coelestibus*, at the instance of the Cardinal of Capua, and of the Bishop of Culma; and, since he undertook the task at the order of Pope Leo X., he dedicated the work to his successor Paul III., and it was received by the Holy Church and studied by all the world.

****

The doctrine of the movement of the earth and the fixity of the sun is condemned on the ground that the Scriptures speak in many places of the sun moving and the earth standing still. The Scriptures not being capable of lying or erring, it followeth that the position of those is erroneous and heretical who maintain that the sun is fixed and the earth in motion.

It is piously spoken that the Scriptures cannot lie. But none will deny that they are frequently abstruse and their true meaning difficult to discover, and more than the bare words signify. One taking the sense too literally might pervert the truth and conceive blasphemies, and give God feet, and hands, and eyes, and human affections, such as anger, repentance, forgetfulness, ignorance, whereas these expressions are employed merely to accommodate the truth to the mental capacity of the unlearned.

This being granted, I think that in the discussion of natural problems we ought to begin not with the Scriptures, but with experiments and demonstrations. Nor does God less admirably discover Himself to us in nature than in Scripture, and having found the truth in nature we may use it as an aid to the true exposition of the Scriptures. The Scriptures were intended to teach men those things which cannot be learned otherwise than by the mouth of the Holy Spirit; but we are meant to use our senses and reason in discovering for ourselves things within their scope and capacity, and hence certain sciences are neglected in the Holy Writ.

Astronomy, for instance, is hardly mentioned, and only the sun, and the moon, and Lucifer are named. Surely, if the holy writers had intended us to derive our astronomical knowledge from the Sacred Books, they would not have left us so uninformed. That they intentionally forbore to speak of the movements and constitution of the stars is the opinion of the most holy and most learned fathers. And if the Holy Spirit has omitted to teach us those matters as not pertinent to our salvation, how can it be said that one view is *de Fide* and the other heretical? I might here insert the opinion of an ecclesiastic raised to the degree of Eminentissimo: That the intention of the Holy Ghost is to teach us how we shall go to Heaven, and not how the heavens go. [Cardinal Baronius]

II. Scriptual and Experimental Truth

Since the Holy Writ is true, and all truth agrees with truth, the truth of Holy Writ cannot be contrary to the truth obtained by reason and experiment. This being true, it is the business of the judicious expositor to find the true meaning of scriptural passages which must accord with the conclusions of observation and experiment, and care must be taken that the work of exposition do not fall into foolish and ignorant hands. It must be remembered that there are very few men capable of understanding both the sacred Scriptures and science, and that there are many with a superficial knowledge of the Scriptures and with no knowledge of science who would fain arrogate to themselves the power of decreeing upon all questions of nature.

Again, to command the professors of astronomy to confute their own observations is to enjoin an impossibility, for it is to command them not to see what they do see, and not to understand what they do understand, and to find what they do not discover. I would entreat the wise and prudent fathers to consider the difference between matters of opinion and matters of demonstration, for demonstrated conclusions touching the things of nature and of the heavens cannot be changed with the same facility as opinions touching what is lawful in a contract, bargain, or bill of exchange. Your highness knows what happened to the late professor of mathematics in the University of Pisa—how, believing that the Copernican doctrine was false, he started to confute it, but in his study became convinced of its truth.

In order to suppress the Copernican doctrine, it would be necessary not only to prohibit the book of Copernicus and the writings of authors who agree with him, but to interdict the whole science of astronomy, and even to forbid men to look at the sky lest they might see Mars and Venus at very varying distances from the earth, and discover Venus at one time crescent, at another time round, or make other observations irreconcilable with the Ptolemaic system.

It is surely harmful to souls to make it a heresy to believe what is proved. The prohibition of astronomy would be an open contempt of a hundred texts of the Holy Scriptures, which teach us that the glory and the greatness of Almighty God are admirably discerned in all His works, and divinely read in the open book of the heavens.

III. Faith and Fact

It may be said that the doctrine of the movement of the sun and the fixity of the earth must *de Fide* be held for true since the Scriptures affirm it, and all the fathers unanimously accept the scriptural words in their naked and literal sense. But it was necessary to assign motion to the sun and rest to the earth lest the shallow minds of the vulgar should be confounded, amused, and rendered obstinate and contumacious with regard to doctrines of faith. St. Jerome writes: "It is the custom for the pen-men of Scripture to deliver their judgments in many things according to the common received opinion that their times had of them." Even Copernicus himself, knowing the power of custom, and unwilling to create confusion in our comprehension, continues to talk of the rising and setting of the sun and stars and of variations in the obliquity of the zodiac. Whence it is to be noted how necessary it is to accommodate our discourse to our accustomed manner of understanding.

In the next place, the common consent of the fathers to a natural proposition should authorise it only if it have been discussed and debated with all possible diligence, and this question was in those times totally buried.

Besides, it is not enough to say that the fathers accept the Ptolemaic doctrine; it is necessary to prove that they condemned the Copernican. Was the Copernican doctrine ever formally condemned as contrary to the Scriptures? And Didacus, discoursing on the Copernican hypothesis, concludes that the motion of the earth is not contrary to the Scriptures.

Let my opponents, therefore, apply themselves to examine the arguments of Copernicus and others; and let them not hope to find such rash and impetuous decisions in the wary and holy fathers, or in the absolute wisdom of him that cannot err, as those into which they have suffered themselves to be hurried by prejudice or personal feeling. His holiness has certainly an absolute power of admitting or condemning propositions not directly *de Fide*, but it is not in the power of any creature to make them true or false otherwise than of their own nature and *de facto* they are.

In my judgment it would be well first to examine the truth of the fact (over which none hath power) before invoking supreme authority; for if it be not possible that a conclusion should be declared heretical while we are not certain but that it may be true, their pains are vain who pretend to condemn the doctrine of the mobility of the earth and the fixity of the sun, unless they have first demonstrated the doctrine to be impossible and false.

—2—
John Henry Newman, "Christianity and Scientific Investigation," from *The Idea of a University* (1858)

Born in London in 1801, John Henry Newman studied at Trinity College, Oxford, and then became a fellow of Oriel College and took Anglican orders. He was appointed vicar of the university church, St. Mary the Virgin, a post he resigned in 1832. Following a trip to the Continent Newman assumed a leading role in the "Oxford movement," a group of high church Anglicans who sought in "Tracts for the Times" to demonstrate the common roots of Roman Catholicism and the Church of England. Newman's study of the issue of apostolic succession led to his conversion to Catholicism in 1845. He spent time in Rome and then returned to England where he helped to found the London Oratory. Newman went to Dublin in 1854 to assume the position of rector of the new Catholic university, where he expounded his philosophy of education in *The Idea of a University* (1858). Of particular relevance in the present context is his conviction that theology and science, when properly practiced, are complementary disciplines.

7. Christianity and Physical Science

But if there be so substantial a truth even in this very broad statement concerning the independence of the fields of Theology and general Science severally, and the consequent impossibility of collision between them, how much more true is that statement, from the very nature of the case, when we contrast Theology, not with Science generally, but definitely with Physics! In Physics is comprised that family of sciences which is concerned with the sensible world, with the phenomena which we see, hear, and handle, or, in other words, with matter. It is the philosophy of matter. Its basis of operations, what it starts from, what it falls back upon, is the phenomena which meet the senses. Those phenomena it ascertains, catalogues, compares, combines, arranges, and then uses for determining something beyond themselves, viz., the order to which they are subservient, or what we commonly call the laws of nature. It never travels beyond the examination of cause and effect. Its object is to resolve the complexity of phenomena into simple elements and principles; but when it has reached those first elements, principles, and laws, its mission is at an end; it keeps within that material system with which it began, and never ventures beyond the *flammantia moenia mundi*. It may, indeed, if it chooses, feel a doubt of the completeness of its analysis hitherto, and for that reason endeavour to arrive at more simple laws and fewer principles. It may be dissatisfied with its own combinations, hypotheses, systems; and leave Ptolemy for Newton, the alchemists for Lavoisier and Davy;—that is, it may decide that it has not yet touched the bottom of its own subject; but still its aim will be to get to the bottom, and nothing more. With matter it began, with matter it {433} will end; it will never trespass into the province of mind. The Hindoo notion is said to believe that the earth stands upon a tortoise; but the physicist, as such, will never ask himself by what influence, external to the universe, the universe is sustained; simply because he is a physicist.

****

Such is Physical Science, and Theology, as is obvious, is just what such Science is not. Theology begins, as its name denotes, not with any sensible facts, phenomena, or results, not with nature at all, but with the Author of nature,—with the one invisible, unapproachable Cause and Source of all things. It begins at the other end of knowledge, and is occupied, not with the finite, but the Infinite. It unfolds and systematizes what He Himself has told us of Himself; of His nature, His attributes, His will, and His acts. As far as it approaches towards Physics, it takes just the counterpart of the questions which occupy the Physical Philosopher. He contemplates facts before him; the Theologian gives the reasons of those facts. The Physicist treats of efficient causes; the Theologian of final. The Physicist tells us of laws; the Theologian of the Author, Maintainer, and Controller of them; of their scope, of their suspension, if so be; of their beginning and their end. This is how the two schools stand related to each other, at that point where they approach the nearest; but for the most part they are absolutely divergent. What Physical Science is engaged in I have already said; as to Theology, it contemplates the world, not of matter, but of mind; the Supreme Intelligence; souls and their destiny; conscience and duty; the past, present, and future dealings of the Creator with the creature. {435}

So far, then, as these remarks have gone, Theology and Physics cannot touch each other, have no intercommunion, have no ground of difference or agreement, of jealousy or of sympathy. As well may musical truths be said to interfere with the doctrines of architectural science; as well may there be a collision between the mechanist and the geologist, the engineer and the grammarian; as well might the British Parliament or the French nation be jealous of some possible belligerent power upon the surface of the moon, as Physics pick a quarrel with Theology.

8. Christianity and Scientific Investigation

Now, let me call your attention, Gentlemen, to what I would infer from these familiar facts. It is, to urge you with an argument à fortiori: viz., that, as you exercise so much exemplary patience in the case of the inexplicable truths which surround so many departments of knowledge, human and divine, viewed in themselves; as you are not at once indignant, censorious, suspicious, difficult of belief, on finding that in the secular sciences one truth is incompatible (according to our human intellect) with another or inconsistent with itself; so you {465} should not think it very hard to be told that there exists, here and there, not an inextricable difficulty, not an astounding contrariety, not (much less) a contradiction as to clear facts, between Revelation and Nature; but a hitch, an obscurity, a divergence of tendency, a temporary antagonism, a difference of tone, between the two,—that is, between Catholic opinion on the one hand, and astronomy, or geology, or physiology, or ethnology, or political economy, or history, or antiquities, on the other. I say that, as we admit, because we are Catholics, that the Divine Unity contains in it attributes, which, to our finite minds, appear in partial contrariety with each other; as we admit that, in His revealed Nature are things, which, though not opposed to Reason, are infinitely strange to the Imagination; as in His works we can neither reject nor admit the ideas of space, and of time, and the necessary properties of lines, without intellectual distress, or even torture; really, Gentlemen, I am making no outrageous request, when, in the name of a University, I ask religious writers, jurists, economists, physiologists, chemists, geologists, and historians, to go on quietly, and in a neighbourly way, in their own respective lines of speculation, research, and experiment, with full faith in the consistency of that multiform truth, which they share between them, in a generous confidence that they will be ultimately consistent, one and all, in their combined results, though there may be momentary collisions, awkward appearances, and many forebodings and prophecies of contrariety, and at all times things hard to the Imagination, though not, I repeat, to the Reason. It surely is not asking them a great deal to beg of them,—since they are forced to admit mysteries in the truths of Revelation, taken by themselves, and in the truths of Reason, taken by themselves,— {466} to beg of them, I say, to keep the peace, to live in good will, and to exercise equanimity, if, when Nature and Revelation are compared with each other, there be, as I have said, discrepancies,—not in the issue, but in the reasonings, the circumstances, the associations, the anticipations, the accidents, proper to their respective teachings.

It is most necessary to insist seriously and energetically on this point, for the sake of Protestants, for they have very strange notions about us. In spite of the testimony of history the other way, they think that the Church has no other method of putting down error than the arm of force, or the prohibition of inquiry. They defy us to set up and carry on a School of Science. For their sake, then, I am led to enlarge upon the subject here. I say, then, he who believes Revelation with that absolute faith which is the prerogative of a Catholic, is not the nervous creature who startles at every sudden sound, and is fluttered by every strange or novel appearance which meets his eyes. He has no sort of apprehension, he laughs at the idea, that any thing can be discovered by any other scientific method, which can contradict any one of the dogmas of his religion. He knows full well there is no science whatever, but, in the course of its extension, runs the risk of infringing, without any meaning of offence on its own part, the path of other sciences: and he knows also that, if there be any one science which, from its sovereign and unassailable position can calmly bear such unintentional collisions on the part of the children of earth, it is Theology. He is sure, and nothing shall make him doubt, that, if anything seems to be proved by astronomer, or geologist, or chronologist, or antiquarian, or ethnologist, in contradiction to the dogmas of faith, that point will eventually turn out, first, {467} not to be proved, or, secondly, not contradictory, or thirdly, not contradictory to any thing really revealed, but to something which has been confused with revelation.

To one notorious instance indeed it is obvious to allude here. When the Copernican system first made progress, what religious man would not have been tempted to uneasiness, or at least fear of scandal, from the seeming contradiction which it involved to some authoritative tradition of the Church and the declaration of Scripture? It was generally received, as if the Apostles had expressly delivered it both orally and in writing, as a truth of Revelation, that the earth was stationary, and that {468} the sun, fixed in a solid firmament, whirled round the earth. After a little time, however, and on full consideration, it was found that the Church had decided next to nothing on questions such as these, and that Physical Science might range in this sphere of thought almost at will, without fear of encountering the decisions of ecclesiastical authority. Now, besides the relief which it afforded to Catholics to find that they were to be spared this addition, on the side of Cosmology, to their many controversies already existing, there is something of an argument in this very circumstance in behalf of the divinity of their Religion. For it surely is a very remarkable fact, considering how widely and how long one certain interpretation of these physical statements in Scripture had been received by Catholics, that the Church should not have formally acknowledged it. Looking at the matter in a human point of view, it was inevitable that she should have made that opinion her own. But now we find, on ascertaining where we stand, in the face of the new sciences of these latter times, that in spite of the bountiful comments which from the first she has ever been making on the sacred text, as it is her duty and her right to do, nevertheless, she has never been led formally to explain the texts in question, or to give them an authoritative sense which modern science may question.

****

On the other hand, it must be of course remembered, Gentlemen, that I am supposing all along good faith, honest intentions, a loyal Catholic spirit, and a deep sense of responsibility. I am supposing, in the scientific inquirer, a due fear of giving scandal, of seeming to countenance views which he does not really countenance, and of siding with parties from whom he heartily differs. I am supposing that he is fully alive to the existence and the power of the infidelity of the age; that he keeps in mind the moral weakness and the intellectual confusion of the majority of men; and that he has no wish at all that any one soul should get harm from certain speculations today, though he may have the satisfaction of being sure that those speculations will, as far as they are erroneous or misunderstood, be corrected in the course of the next half-century.

—3—
John A. Zahm, *Evolution and Dogma,* Chicago: D. H. Mcbride and Co., 1896; Rpt. Ed. Arno Press, 1978

Holy Cross priest John Zahm (1851–1921) was professor of physics and chemistry at the University of Notre Dame, and became one of the foremost American apologists for the harmonization of evolutionary theory with Catholic dogma. Zahm's book attracted the attention of the Roman curia in part because of the existing campaign against "Americanism." The decree of condemnation was never published by the Congregation of the Index, and Zahm never issued a retraction. A large part of *Evolution and Dogma* examines Darwin's theory of evolution from the perspective of various sciences. In this selection, Zahm considers its relation to Christian dogma, and argues that not only does faith have nothing to fear from it but that evolution is in fact an ennobling conception.

Faith Has Nothing to Apprehend from Evolution

Suppose, then, that a demonstrative proof of the theory of Evolution should eventually be given, a proof such as would satisfy the most exacting and the most skeptical, it is evident, from what has already been stated, that Catholic Dogma would remain absolutely intact and unchanged. Individual theorists would be obliged to accommodate their views to the facts of nature, but the doctrines of the Church would not be affected in the slightest. The hypothesis of St. Augustine and St. Thomas Aquinas would then become a thesis, and all reasonable and consistent men would yield ready, unconditional and unequivocal assent.

And suppose, further, that in the course of time science shall demonstrate—a most highly improbable event—the animal origin of man as to his body. There need, even then, be no anxiety so far as the truths of faith are concerned, Proving that the body of the common ancestor of humanity is descended from some higher form of ape, or from some extinct anthropopithecus, would not necessarily contravene either the declarations of Genesis, or the principles regarding derivative creation which found acceptance with the greatest of the Church's Fathers and Doctors.

Mr. Gladstone, in the work just quoted from, expresses the same idea with characteristic force and lucidity. " If," he says, "while Genesis asserts a separate creation of man, science should eventually prove that man sprang, by a countless multitude of indefinitely small variations, from a lower, and even from the lowest ancestry, the statement of the great chapter would still remain undisturbed. For every one of those variations, however minute, is absolutely separate, in the points wherein it varies, from what followed and also from what preceded it; is in fact and in effect a distinct or separate creation. And the fact that the variation is so small that, taken singly, our use may not be to reckon it, is nothing whatever to the purpose. For it is the finiteness of our faculties which shuts us off by a barrier downward, beyond a certain limit, from the small, as it shuts us off by a barrier upward from the great; whereas for Him whose faculties are infinite, the small and the great are, like the light and the darkness, 'both alike,' and if man came up by innumerable stages from a low origin to the image of God, it is God only who can say, as He has said in other cases, which of those stages may be worthy to be noted with the distinctive name of creation, and at what point of the ascent man could first be justly said to exhibit the image of God."

But the derivation of man from the ape, we are told, degrades man. Not at all. It would be truer to say that such derivation ennobles the ape. Sentiment aside, it is quite unimportant to the Christian "whether he is to trace back his pedigree directly or indirectly to the dust." St. Francis of Assisi, as we learn from his life, "called the birds his brothers." Whether he was correct, either theologically or zoologically, he was plainly free from that fear of being mistaken for an ape which haunts so many in these modern times. Perfectly sure that he, himself, was a spiritual being, he thought it at least possible that birds might be spiritual beings, likewise incarnate like himself in mortal flesh; and saw no degradation to the dignity of human nature in claiming kindred lovingly with creatures so beautiful, so wonderful, who, as he fancied, "praised God in the forest, even as angels did in heaven" (Kingsley, Prose Idylls., 24ff).

Misapprehensions Regarding Evolution

Many, it may here be observed, look on the theory of Evolution with suspicion, because they fail to understand its true significance. They seem to think that it is an attempt to account for the origin of things when, in reality, it deals only with their historical development. It deals not with creation, with the origin of things, but with the *modus creandi*, or, rather, with the *modus formandi*, after the universe was called into existence by Divine Omnipotence. Evolution, then, postulates creation as an intellectual necessity, for if there had not been a creation there would have been nothing to evolve, and Evolution would, therefore, have been an impossibility.

And for the same reason, Evolution postulates and must postulate, a Creator, the sovereign Lord of all things, the Cause of causes, the *terminus a quo* as well as the *terminus ad quern* of all that exists or can exist. But Evolution postulates still more. In order that Evolution might be at all possible it was necessary that there should have been not only an antecedent creation *ex nihilo,* but also that there should have been an antecedent involution, or a creation *in potentia.* To suppose that simple brute matter could, by its own motion or by any power inherent in matter as such, have been the sole efficient cause of the Evolution of organic from inorganic matter, of the higher from the lower forms of life, of the rational from the irrational creature, is to suppose that a thing can give what it does not possess, that the greater is contained in the less, the superior in the inferior, the whole in a part.

No mere mechanical theory, therefore, however ingenious, is competent to explain the simplest fact of development. Not only is such a theory unable to account for the origin of a speck of protoplasm, or the germination of a seed, but it is equally incompetent to assign a reason for the formation of the smallest crystal or the simplest chemical compound. Hence, to be philosophically valid, Evolution must postulate a Creator not only for the material which is evolved, but it must also postulate a Creator, *Causa causarum,* for the power or agency which makes any development possible. God, then, not only created matter in the beginning, but He gave it the power of evolving into all forms it has since assumed or ever shall assume.

But this is not all. In order to have an intelligible theory of Evolution, a theory that can meet the exacting demands of a sound philosophy as well as of a true theology, still another postulate is necessary. We must hold not only that there was an actual creation of matter in the beginning, that there was a potential creation which rendered matter capable of Evolution, in accordance with the laws impressed by God on matter, but we must also believe that creative action and influence still persist, that they always have persisted from the dawn of creation, that they, and they alone, have been efficient in all the countless stages of evolutionary progress from atoms to monads, from monads to man.

This ever-present action of the Deity, this immanence of His in the work of His hands, this continuing in existence and developing of the creatures He has made, is what St. Thomas calls the "Divine administration," and what is ordinarily known as Providence. It connotes the active and constant cooperation of the Creator with the creature, and implies that if the multitudinous forms of terrestrial life have been evolved from the potentiality of matter, they have been so evolved because matter was in the first instance proximately disposed for Evolution by God Himself, and has ever remained so disposed. To say that God created the universe in the beginning, and that He gave matter the power of developing into all the myriad forms it subsequently exhibited, but that after doing this He had no further care for what He had brought into existence, would be equivalent to indorsing the Deism of Hume, or to affirming the old pagan notion according to which God, after creating the world, withdrew from it and left it to itself.

Well, then, can we say of Evolution what Dr. Martineau says of science, that it "discloses the method of the world, not its cause; religion, its cause and not its method" (Martineau, 1891). Evolution is the grand and stately march of creative energy, the sublime manifestation of what Claude Bernard calls "the first, creative, legislative and directing Cause." In it we have constantly before our eyes the daily miracles, *quotidiana Dei miracula,* of which St. Augustine speaks, and through it we are vouchsafed a glimpse, as it were, of the operation of Providence in the government of the world.

Evolution, therefore, is neither a " philosophy of mud," nor " a gospel of dirt," as it has been denominated. So far, indeed, is this from being the case that, when properly understood, it is found to be a strong and useful ally of Catholic Dogma. For if Evolution be true, the existence of God and an original creation follow as necessary inferences. "A true development," as has truthfully been asserted, "implies a *terminus a quo* as well as a *terminus ad quem.* If, then, Evolution is true, an absolute beginning, however unthinkable, is probable;"— I should say certain—"the eternity of matter is inconsistent with scientific Evolution" (Moore, 229).

"Nature," Pascal somewhere says, "confounds the Pyrrhonist, and reason, the dogmatist." Evolution, we can declare with equal truth, confounds the agnostic, and science, the atheist. For, as an English positivist has observed: "You cannot make the slightest concession to metaphysics without ending in a theology," a statement which is tantamount to the admission that "If once you allow yourself to think of the origin and end of things, you will have to believe in a God." And the God you will have to believe in is not an abstract God, an unknowable xn, a mere metaphysical deity, "defecated to a pure transparency," but a personal God, a merciful and loving Father.

As to man, Evolution, far from depriving him of his high estate, confirms him in it, and that, too, by the strongest and noblest of titles. It recognizes that although descended from humble lineage, he is "the beauty of the world, and the paragon of animals; "that although from dust— tracing his lineage back to its first beginnings—he is of the "quintessence of dust." It teaches, and in the most eloquent language, that he is the highest term of a long and majestic development, and replaces him "in his old position of headship in the universe, even as in the days of Dante and Aquinas."

Evolution an Ennobling Conception

And as Evolution ennobles our conceptions of God and of man, so also does it permit us to detect new beauties, and discover new lessons, in a world that, according to the agnostic and monistic views, is so dark and hopeless. To the one who says there is no God, "the immeasurable universe," in the language of Jean Paul, "has become but a cold mass of iron, which hides an eternity without form and void."

To the theistic evolutionist, however, all is instinct with invitations to a higher life and a happier existence in the future; all is vocal with hymns of praise and benediction. Everything is a part of a grand unity betokening an omnipotent Creator. All is foresight, purpose, wisdom. We have the entire history of the world and of all systems of worlds, "gathered, as it were, into one original, creative act, from which the infinite variety of the universe has come, and more is coming yet" (Temple, 116). And God's hand is seen in the least as in the greatest. His power and goodness are disclosed in the beauteous crystalline form of the snow-flake, in the delicate texture, fragrance and color of the rose, in the marvelous pencilings of the butterfly's wing, in the gladsome and melodious notes of the lark and the thrush, in the tiniest morning dew-drop with all its gorgeous prismatic hues and wondrous hidden mysteries. All are pregnant with truths of the highest order, and calculated to inspire courage, and to strengthen our hope in faith's promise of a blissful immortality.

The Divine it is which holds all things together (Aristotle, 1999: Metaphysics, XI, viii). So taught the old Greek philosophy as reported by the most gifted of her votaries. And this teaching of the sages of days long past, is extended and illuminated by the far-reaching generalization of Evolution, in a manner that is daily becoming more evident and remarkable. But what Greek philosophy faintly discerned, and what Evolution distinctly enunciates, is rendered gloriously manifest by the declaration of revealed truth, and by the doctrines of Him who is the Light of the World.

Science and Evolution tell us of the transcendence and immanence of the First Cause, of the Cause of causes, the Author of all the order and beauty in the world, but it is revelation which furnishes us with the strongest evidence of the relations between the natural and supernatural orders, and brings out in the boldest relief the absolute dependence of the creature on its Maker. It is faith which teaches us how God "binds all together into Himself;" how He quickens and sustains "each thing separately, and all as collected in one."

I can, indeed, no better express the ideas which Evolution so beautifully shadows forth, nor can I more happily conclude this long discussion than by appropriating the words used long ago by that noble champion of the faith, St. Athanasius. "As the musician," says the great Alexandrine Doctor, in his "*Oratio Contra Gentiles*," "having tuned his lyre, and harmonized together the high with the low notes, and the middle notes with the extremes, makes the resulting music one; so the Wisdom of God, grasping the universe like a lyre, blending the things of air with those of earth, and the things of heaven with those of air, binding together the whole and the parts, and ordering all by His counsel and His will, makes the world itself and its appointed order one in fair and harmonious perfection; yet He, Himself, moving all things, remains unmoved with the Father" (Athanasius, *Oratio Contra Gentiles*, XLII).

— 4 —
Pope Pius XII, *Humani Generis* Aug. 12, 1950; Sections 5–9; 11–14; 35–37

Copyright Holder: Libreria Edificie Vaticana

In this apologetic encyclical, Pope Pius XII distinguishes between what he sees as the erroneous, materialistic interpretations of evolution on the one hand and the scientific credibility of evolutionary theory on the other hand. Pius XII also draws on the distinction between the body and soul in order to justify why, for Catholics, it is possible to accept the evolutionary origins of the human species.

5. If anyone examines the state of affairs outside the Christian fold, he will easily discover the principle trends that not a few learned men are following. Some imprudently and indiscreetly hold that evolution, which has not been fully proved even in the domain of natural sciences, explains the origin of all things, and audaciously support the monistic and pantheistic opinion that the world is in continual evolution. Communists gladly subscribe to this opinion so that, when the souls of men have been deprived of every idea of a personal God, they may the more efficaciously defend and propagate their dialectical materialism.

6. Such fictitious tenets of evolution which repudiate all that is absolute, firm and immutable, have paved the way for the new erroneous philosophy which, rivaling idealism, immanentism and pragmatism, has assumed the name of existentialism, since it concerns itself only with existence of individual things and neglects all consideration of their immutable essences.

7. There is also a certain historicism, which attributing value only to the events of man's life, overthrows the foundation of all truth and absolute law, both on the level of philosophical speculations and especially to Christian dogmas.

8. In all this confusion of opinion it is some consolation to Us to see former adherents of rationalism today frequently desiring to return to the fountain of divinely communicated truth, and to acknowledge and profess the word of God as contained in Sacred Scripture as the foundation of religious teaching. But at the same time it is a matter of regret that not a few of these, the more firmly they accept the word of God, so much the more do they diminish the value of human reason, and the more they exalt the authority of God the Revealer, the more severely do they spurn the teaching office of the Church, which has been instituted by Christ, Our Lord, to preserve and interpret divine revelation. This attitude is not only plainly at variance with Holy Scripture, but is shown to be false by experience also. For often those who disagree with the true Church complain openly of their disagreement in matters of dogma and thus unwillingly bear witness to the necessity of a living Teaching Authority.

9. Now Catholic theologians and philosophers, whose grave duty it is to defend natural and supernatural truth and instill it in the hearts of men, cannot afford to ignore or neglect these more or less erroneous opinions. Rather they must come to understand these same theories well, both because diseases are not properly treated unless they are rightly diagnosed, and because sometimes even in these false theories a certain amount of truth is contained, and, finally, because these theories provoke more subtle discussion and evaluation of philosophical and theological truths.

...

11. Another danger is perceived which is all the more serious because it is more concealed beneath the mask of virtue. There are many who, deploring disagreement among men and intellectual confusion, through an imprudent zeal for souls, are urged by a great and ardent desire to do away with the barrier that divides good and honest men; these advocate an "eirenism" according to which, by setting aside the questions which divide men, they aim not only at joining forces to repel the attacks of atheism, but also at reconciling things opposed to one another in the field of dogma. And as in former times some questioned whether the traditional apologetics of the Church did not constitute an obstacle rather than a help to the winning of souls for Christ, so today some are presumptive enough to question seriously whether theology and theological methods, such as with the approval of ecclesiastical authority are found in our

schools, should not only be perfected, but also completely reformed, in order to promote the more efficacious propagation of the kingdom of Christ everywhere throughout the world among men of every culture and religious opinion.

12. Now if these only aimed at adapting ecclesiastical teaching and methods to modern conditions and requirements, through the introduction of some new explanations, there would be scarcely any reason for alarm. But some through enthusiasm for an imprudent "eirenism" seem to consider as an obstacle to the restoration of fraternal union, things founded on the laws and principles given by Christ and likewise on institutions founded by Him, or which are the defense and support of the integrity of the faith, and the removal of which would bring about the union of all, but only to their destruction.

13. These new opinions, whether they originate from a reprehensible desire of novelty or from a laudable motive, are not always advanced in the same degree, with equal clarity nor in the same terms, nor always with unanimous agreement of their authors. Theories that today are put forward rather covertly by some, not without cautions and distinctions, tomorrow are openly and without moderation proclaimed by others more audacious, causing scandal to many, especially among the young clergy and to the detriment of ecclesiastical authority. Though they are usually more cautious in their published works, they express themselves more openly in their writings intended for private circulation and in conferences and lectures. Moreover, these opinions are disseminated not only among members of the clergy and in seminaries and religious institutions, but also among the laity, and especially among those who are engaged in teaching youth.

14. In theology some want to reduce to a minimum the meaning of dogmas; and to free dogma itself from terminology long established in the Church and from philosophical concepts held by Catholic teachers, to bring about a return in the explanation of Catholic doctrine to the way of speaking used in Holy Scripture and by the Fathers of the Church. They cherish the hope that when dogma is stripped of the elements which they hold to be extrinsic to divine revelation, it will compare advantageously with the dogmatic opinions of those who are separated from the unity of the Church and that in this way they will gradually arrive at a mutual assimilation of Catholic dogma with the tenets of the dissidents.

...

35. It remains for Us now to speak about those questions which, although they pertain to the positive sciences, are nevertheless more or less connected with the truths of the Christian faith. In fact, not a few insistently demand that the Catholic religion take these sciences into account

as much as possible. This certainly would be praiseworthy in the case of clearly proved facts; but caution must be used when there is rather question of hypotheses, having some sort of scientific foundation, in which the doctrine contained in Sacred Scripture or in Tradition is involved. If such conjectural opinions are directly or indirectly opposed to the doctrine revealed by God, then the demand that they be recognized can in no way be admitted.

36. For these reasons the Teaching Authority of the Church does not forbid that, in conformity with the present state of human sciences and sacred theology, research and discussions, on the part of men experienced in both fields, take place with regard to the doctrine of evolution, in as far as it inquires into the origin of the human body as coming from preexistent and living matter—for the Catholic faith obliges us to hold that souls are immediately created by God. However, this must be done in such a way that the reasons for both opinions, that is, those favorable and those unfavorable to evolution, be weighed and judged with the necessary seriousness, moderation and measure, and provided that all are prepared to submit to the judgment of the Church, to whom Christ has given the mission of interpreting authentically the Sacred Scriptures and of defending the dogmas of faith.[11] Some however, rashly transgress this liberty of discussion, when they act as if the origin of the human body from pre-existing and living matter were already completely certain and proved by the facts which have been discovered up to now and by reasoning on those facts, and as if there were nothing in the sources of divine revelation which demands the greatest moderation and caution in this question.

37. When, however, there is question of another conjectural opinion, namely polygenism, the children of the Church by no means enjoy such liberty. For the faithful cannot embrace that opinion which maintains that either after Adam there existed on this earth true men who did not take their origin through natural generation from him as from the first parent of all, or that Adam represents a certain number of first parents. Now it is no way apparent how such an opinion can be reconciled with that which the sources of revealed truth and the documents of the Teaching Authority of the Church propose with regard to original sin, which proceeds from a sin actually committed by an individual Adam and which, through generation, is passed on to all and is in everyone as his own.[12]

NOTES

11. Cfr. Allocut Pont. to the members of the Academy of Science, November 30, 1941: A.A.S., vol. XXXIII, p. 506.

12. Cfr. *Rom.*, V, 12–19; Conc. Trid., sess, V, can. 1–4.

—5—
"Discourse of His Holiness Pope John Paul II on October 28, 1986 at the Solemn Audience Granted in Occasion of the Fiftieth Anniversary of the Pontifical Academy of Sciences," in Martini-Bettòlo, G.B. *Discourses of the Popes from Pius XI to John Paul II to the Pontifical Academy of Sciences 1936–1986* (Pontificia Academia Scientiarum, 1986), Sections 1–6, 193–197

In this 1986 speech, Pope John Paul II acknowledges the fiftieth anniversary of the renewal of the 'Linceo', what is now called The Pontifical Academy of Sciences. This speech contains themes characteristic of John Paul II's papacy, including regret over the Galileo episode, the significance of science's pursuit of truth, and reference to the practical, cultural dimensions of science.

Discourse of His Holiness Pope John Paul II given on 28th October 1986 at the Solemn Audience granted in occasion of the fiftieth Anniversary of the Pontifical Academy of Sciences.

Your Eminences,
Mr. Director-General of UNESCO,
Mr. Minister of Scientific Research in Italy,
Your Excellencies,
Ladies and Gentlemen,

It is with great joy that I celebrate with you the fiftieth anniversary of the act by which Pope Pius XI renewed the Pontifical Academy of the "new Lincei" and made it the *Pontifical Academy of Sciences* with the Motu proprio *In multis solaciis* of 28 October 1936.

1. The word "Linceo" belongs to your history and to your very being, dear Academicians, because you draw your origin and your fundamental inspiration from the group of young scientists who were gathered by Prince Federico Cesi and gave birth in the year 1603 to the Academy of the "Lincei"; Galileo Galilei became a member in the year 1611 and thereafter signed all his works with the title "Linceo."

The bonds between the Church and the Academy became particularly intense under Pius IX, who entrusted to it tasks of scientific research in the service of the Papal States, and the relationship became even deeper under his successors, especially under Pius XI, who conferred on it the title and the function of *scientific Senate* of the Church, made up of seventy members whom the Sovereign Pontiff asked to "promote ever more and ever better the progresses of the sciences," adding: "We do not ask anything more of them, for this noble goal and this sublime task constitute the service that we expect from men closely bound to the truth."

My venerated predecessors Pius XII, John XXIII and Paul VI encouraged the Pontifical Academy, fully convinced of the indispensable role of science in the service of created truth, and ultimately in the service of the First Truth, who is God, following the path from the finite to the infinite-a path that is printed on the human spirit.

The Sovereign Pontiffs were actively supported in this by the succession of Presidents, from Father Agostino Gemelli, Monsignor Georges Lemaître, and Father Daniel O'Connell to Professor Carlos Chagas, whom I thank warmly for the important work which he has carried out. Thanks to these Presidents, thanks also to the collaboration of all the members of the Chancellery, this Academy has acquired a celebrated prestige and a scientific role on a very high level, awakening elsewhere participation in important work of many representatives of the world scientific community.

2. In the course of the fifty years of your history, Ladies and Gentlemen, you have very properly given primacy to *pure* science, claiming for it its legitimate autonomy. When I addressed you in my first discourse in this very place, on 10 November 1979, I proclaimed the dignity and the high value of science, with regard to its theoretical side: "Fundamental research must be free in its relationship to political and economic power, which must cooperate in its development, without placing obstacles in its path ... Like every other truth, scientific truth is obliged to give account of itself only to itself and to the supreme truth that is God, the creator of man and of everything."

In addition to pure science, you have dedicated yourselves to the study of its consequences for *applied sciences*, which—as I said in that same discourse— "has rendered and will render immense services to man, provided that it is inspired by love, guided by wisdom, and accompanied by the courage that defends it against the undue interference of all tyrannic powers." Your Academy has been actively involved in the applied sciences as these relate to the needs of humanity as a whole, always in awareness of the requirements of the moral law.

3. The existence and the activity of this Academy, which was founded by the Holy See and is in constant liaison with it, illustrate above all the fact that there is *no contradiction between science and religion. The Church* esteems science, and even recognises a certain connaturality with those who dedicate their endeavours to science, as with all who seek to open

up the human family to the noblest values of the true, the good and the beautiful, to the understanding of the things that have universal value (cf. *Gaudium et Spes,* 57, par. 3). The Pontifical Academy, for its part, shows clearly that *science* likewise needs to be in harmony with wisdom and with ethics, in order to satisfy the deepest requirements of man's spirit and heart, so that his dignity may be safeguarded.

A new type of dialogue has now begun between the Church and the world of science. In my discourse to men of science and students at Cologne, on 15 November 1980, I went so far as to say: "The Church takes up the defense of reason and of science, recognising that science has the capacity to attain to the truth... defending the freedom of science which gives it its dignity as a human and personal good..." If divergences can appear between the Church and science, "the reason for this must be sought in the finitude of our reason, which is limited in its extent and thus exposed to error."

4. We have the good fortune to experience today the close of a history in which the harmony between scientific culture and Christianity was not always easy (cf. *Gaudium et Spes,* 62). At the beginning of this discourse, I recalled the institution which prefigured the Academy around the year 1600. But one must consider above all the manner in which the question of the relationship between theology and the natural sciences was posed on the threshold of the modern period.

Isaac Newton synthesised and brought to completion the discoveries of Kepler, Copernicus, Galileo and Descartes; he was the witness and the decisive agent of *the scientific revolution* of the seventeenth century. It was then that modern science broke through the traditional boundaries which had been determined hitherto by a geocentric view of the universe and by a conception of the elements of nature that was more qualitative than quantitative. These great *scholars* who were experts in an experimental study of the universe, with ever increasing precision and specialisation, did not the less remain in an attitude that sought the global meaning of nature; their speculation as *thinkers* about the cosmos bears witness to this. Their bold researches helped to define better *the boundaries between the different orders of knowledge.* They were not always accepted on this point, and the Church herself took a long time to become reconciled to their points of view.

The experience of *Galileo* is a typical illustration of this. Although it was a painful experience indeed, it rendered an invaluable service to the world of science and to the Church, leading us to understand better the relationship between the revealed Truth and the truths that are discovered empirically. Galileo himself did not accept a genuine contradiction between science and faith: both come from the same Sources and are to be brought into relationship with the first Truth.

Christians have been led to read the Bible afresh, without seeking in it a scientific cosmological system. And scientists themselves have been invited to remain open to the absoluteness of God and to an awareness of creation. In itself, no field is barred to scientific investigation, provided that this respects the human being; it is, rather, the methodologies employed that bring the scientists to make certain abstractions and delimitations.

5. One could mention other very vivid tensions that belong-let us hope-to a vanished past. *In the last century,* in the name of the new sciences and the new philosophies, positivism blamed the traditional positions of the Church, accusing her of being opposed to science and to research. Leo XIII took up the challenge and showed that the Church joyfully welcomes whatever permits man to explore nature better and to improve the human condition. At the same time, he gave a vigorous impulse to the renewal of the ecclesiastical sciences.

In our days, *the distinction* and the complementarity of the orders of knowledge-the order of faith and the order of reason-were expressed with decisive clarity in the teaching of the *Second Vatican Council:* "The Church affirms the legitimate autonomy of culture, and particularly that of the sciences" (*Gaudium et Spes,* 59, par. 3). "It is by virtue of creation itself that all things are established in accordance with their own substance, truth and excellence, with their ordering and their specific laws" (*ibid.,* 36, par. 2). One must recognise the particular methods of each of the sciences. "This is why methodical research, in all the fields of knowledge, will never be truly opposed to faith, if it is carried out in a truly scientific manner and follows the norms of morality: worldly realities and the realities of faith find their origin in the same God" (ibid.). But it would be false to understand this autonomy of earthly realities to mean that they did not depend on God, and that man could dispose of them without reference to the Creator.

The principles are clear, and ought from now onwards to remove every attitude of fear or of defiance, but this does not mean that every difficulty is resolved: new researches and discoveries of the sciences pose *new questions* which will all be new demands for theologians in the way that they present the truths of the faith while always safeguarding the sense and the meaning of these truths (cf. *ibid.,* 62, par. 2). But the scientists themselves for their part go on to make a criticism of their methods and objectives.

Today, far from shutting herself up in an apologetic or defensive perspective, the Church rather makes herself the advocate of science, of reason, and of the freedom of research, to legitimize authentic science. Your Academy can bear witness to this. And I speak here, beyond your own persons, to the scientific community of the whole world.

6. It is indeed important to situate scientific endeavour *within the general context of culture.* Man can never neglect to ask himself the question of the profound meaning of culture and of science for the human person (d. *ibid.,* 61, par. 4).

Man lives a truly human life thanks to culture, that is, by cultivating the goods and the values of nature, affirming and developing the manifold capacities of his spirit and his body. A principal aspect of culture is the submission of the universe by means of knowledge (d. *ibid.,* 5.3). The widening and deepening of scientific knowledge constitute therefore an undeniable progress for man, because this brings him closer to a precise knowledge of the truth.

This free search for truth for its own sake is one of the noblest prerogatives of man. Science goes astray if it ceases to pursue its ultimate end, which is the service of culture and hence of man; it experiences crisis when it is reduced to a purely utilitarian model; it is corrupted when it becomes a technical instrument of domination or manipulation for economic or political goals. There is then what one can call a crisis of the legitimation of science, and it is therefore urgent to defend authentic science that is open to the question of the meaning of man and to the search for the whole truth, a *free science that is dependent only on the truth.* From the point of view of the Church, it would be impossible to separate science and culture.

In the same way, the Church considers man to be not only the object of culture, but its subject, and she encourages the work of science: she appreciates not only the scientists' use of intelligence, but their professional and moral merit, their intellectual honesty, their objectivity their search for what is true, their self-discipline, their cooperation in teams, their commitment to serve man, their respect in the presence of the mysteries of the universe. These are human values that display the spiritual vocation of man.

— 6 —
Pope John Paul II, "Truth Cannot Contradict Truth" Statement to the Pontifical Academy of Sciences, Oct. 22, 1996

Copyright – Libreria Edificie Vaticana and The Pontifical Academy of Sciences, 1987

This is one of the most widely discussed speeches in the papacy of Pope John Paul II. In it, he describes evolution as 'more than a hypothesis.' Based on the framework set out by Pius XII in *Humani*

Generis in 1950, Pope John Paul II discusses the uniqueness of human beings, each made immediately by God.

Truth Cannot Contradict Truth

Address of Pope John Paul II to the Pontifical Academy of Sciences (October 22, 1996)

WITH GREAT PLEASURE I address cordial greeting to you, Mr. President, and to all of you who constitute the Pontifical Academy of Sciences, on the occasion of your plenary assembly. I offer my best wishes in particular to the new academicians, who have come to take part in your work for the first time. I would also like to remember the academicians who died during the past year, whom I commend to the Lord of life.

1. In celebrating the 60th anniversary of the academy's refoundation, I would like to recall the intentions of my predecessor Pius XI, who wished to surround himself with a select group of scholars, relying on them to inform the Holy See in complete freedom about developments in scientific research, and thereby to assist him in his reflections.

He asked those whom he called the Church's "senatus scientificus" to serve the truth. I again extend this same invitation to you today, certain that we will be able to profit from the fruitfulness of a trustful dialogue between the Church and science (cf. Address to the Academy of Sciences, No. 1, Oct. 28, 1986; *L'Osservatore Romano, Eng. ed.*, Nov. 24, 1986, p. 22).

2. I am pleased with the first theme you have chosen, that of the origins of life and evolution, an essential subject which deeply interests the Church, since revelation, for its part, contains teaching concerning the nature and origins of man. How do the conclusions reached by the various scientific disciplines coincide with those contained in the message of revelation? And if, at first sight, there are apparent contradictions, in what direction do we look for their solution? We know, in fact, that truth cannot contradict truth (cf. Leo XIII, encyclical *Providentissimus Deus*). Moreover, to shed greater light on historical truth, your research on the Church's relations with science between the 16th and 18th centuries is of great importance. During this plenary session, you are undertaking a "reflection on science at the dawn of the third millennium," starting with the identification of the principal problems created by the sciences and which affect humanity's future. With this step you point the way to solutions which will be beneficial to the whole human community. In the domain of inanimate and animate nature, the evolution of science and its applications give rise to new questions. The better the Church's knowledge is of their essential aspects, the more she will understand their impact. Consequently, in accordance with her specific mission she will be able to offer criteria for

discerning the moral conduct required of all human beings in view of their integral salvation.

3. Before offering you several reflections that more specifically concern the subject of the origin of life and its evolution, I would like to remind you that the magisterium of the Church has already made pronouncements on these matters within the framework of her own competence. I will cite here two interventions.

In his encyclical *Humani Generis* (1950), my predecessor Pius XII had already stated that there was no opposition between evolution and the doctrine of the faith about man and his vocation, on condition that one did not lose sight of several indisputable points.

For my part, when I received those taking part in your academy's plenary assembly on October 31, 1992, I had the opportunity with regard to Galileo to draw attention to the need of a rigorous hermeneutic for the correct interpretation of the inspired word. It is necessary to determine the proper sense of Scripture, while avoiding any unwarranted interpretations that make it say what it does not intend to say. In order to delineate the field of their own study, the exegete and the theologian must keep informed about the results achieved by the natural sciences (cf. AAS 85 1/81993 3/8, pp. 764–772; address to the Pontifical Biblical Commission, April 23, 1993, announcing the document on the *The Interpretation of the Bible in the Church*: AAS 86 1/81994 3/8, pp. 232–243).

4. Taking into account the state of scientific research at the time as well as of the requirements of theology, the encyclical *Humani Generis* considered the doctrine of "evolutionism" a serious hypothesis, worthy of investigation and in-depth study equal to that of the opposing hypothesis. Pius XII added two methodological conditions: that this opinion should not be adopted as though it were a certain, proven doctrine and as though one could totally prescind from revelation with regard to the questions it raises. He also spelled out the condition on which this opinion would be compatible with the Christian faith, a point to which I will return. Today, almost half a century after the publication of the encyclical, new knowledge has led to the recognition of the theory of evolution as more than a hypothesis. [*Aujourdhui, près dun demi-siècle après la parution de l'encyclique, de nouvelles connaissances conduisent à reconnaitre dans la théorie de l'évolution plus qu'une hypothèse.*] It is indeed remarkable that this theory has been progressively accepted by researchers, following a series of discoveries in various fields of knowledge. The convergence, neither sought nor fabricated, of the results of work that was conducted independently is in itself a significant argument in favor of this theory.

What is the significance of such a theory? To address this question is to enter the field of epistemology. A theory is a metascientific elaboration, distinct from the results of observation but consistent with them. By means of it a series of independent data and facts can be related and interpreted in a unified explanation. A theory's validity depends on whether or not it can be verified; it is constantly tested against the facts; wherever it can no longer explain the latter, it shows its limitations and unsuitability. It must then be rethought.

Furthermore, while the formulation of a theory like that of evolution complies with the need for consistency with the observed data, it borrows certain notions from natural philosophy.

And, to tell the truth, rather than the theory of evolution, we should speak of several theories of evolution. On the one hand, this plurality has to do with the different explanations advanced for the mechanism of evolution, and on the other, with the various philosophies on which it is based. Hence the existence of materialist, reductionist and spiritualist interpretations. What is to be decided here is the true role of philosophy and, beyond it, of theology.

5. The Church's magisterium is directly concerned with the question of evolution, for it involves the conception of man: Revelation teaches us that he was created in the image and likeness of God (cf. Gn 1:27–29). The conciliar constitution *Gaudium et Spes* has magnificently explained this doctrine, which is pivotal to Christian thought. It recalled that man is "the only creature on earth that God has wanted for its own sake" (No. 24). In other terms, the human individual cannot be subordinated as a pure means or a pure instrument, either to the species or to society; he has value *per se*. He is a person. With his intellect and his will, he is capable of forming a relationship of communion, solidarity and self-giving with his peers. St. Thomas observes that man's likeness to God resides especially in his speculative intellect, for his relationship with the object of his knowledge resembles God's relationship with what he has created (Summa Theologica I-II:3:5, ad 1). But even more, man is called to enter into a relationship of knowledge and love with God himself, a relationship which will find its complete fulfillment beyond time, in eternity. All the depth and grandeur of this vocation are revealed to us in the mystery of the risen Christ (cf. *Gaudium et Spes*, 22). It is by virtue of his spiritual soul that the whole person possesses such a dignity even in his body. Pius XII stressed this essential point: If the human body take its origin from pre-existent living matter, the spiritual soul is immediately created by God ("animas enim a Deo immediate creari catholica fides nos retinere iubei"; "Humani Generis," 36). Consequently, theories of evolution which, in accordance with the philosophies inspiring them, consider the spirit as emerging from the forces of living matter or as a mere *epiphenomenon* of this matter, are incompatible with the truth about man. Nor are they able to ground the dignity of the person.

6. With man, then, we find ourselves in the presence of an ontological difference, an ontological leap, one could say. However, does not the posing of such ontological discontinuity run counter to that physical continuity which seems to be the main thread of research into evolution in the field of physics and chemistry? Consideration of the method used in the various branches of knowledge makes it possible to reconcile two points of view which would seem irreconcilable. The sciences of observation describe and measure the multiple manifestations of life with increasing precision and correlate them with the time line. The moment of transition to the spiritual cannot be the object of this kind of observation, which nevertheless can discover at the experimental level a series of very valuable signs indicating what is specific to the human being. But the experience of metaphysical knowledge, of self-awareness and self-reflection, of moral conscience, freedom, or again of aesthetic and religious experience, falls within the competence of philosophical analysis and reflection, while theology brings out its ultimate meaning according to the Creator's plans.

7. In conclusion, I would like to call to mind a Gospel truth which can shed a higher light on the horizon of your research into the origins and unfolding of living matter. The Bible in fact bears an extraordinary message of life. It gives us a wise vision of life inasmuch as it describes the loftiest forms of existence. This vision guided me in the encyclical which I dedicated to respect for human life, and which I called precisely "Evangelium Vitae."

It is significant that in St. John's Gospel *life* refers to the divine light which Christ communicates to us. We are called to enter into eternal life, that is to say, into the eternity of divine beatitude. To warn us against the serious temptations threatening us, our Lord quotes the great saying of Deuteronomy: "Man shall not live by bread alone, but by every word that proceeds from the mouth of God" (Dt 8:3; cf. Mt 4:4). Even more, "life" is one of the most beautiful titles which the Bible attributes to God. He is the living God.

I cordially invoke an abundance of divine blessings upon you and upon all who are close to you.

Excerpted from the October 30 issue of the English edition of L'Osservatore Romano.

—7—
Pope Paul VI, *Humanae Vitae* 1–2; 4–11.

Copyright – Libreria Edificie Vaticana and The Pontifical Academy of Sciences, 1987

This encyclical, issued in 1968, was a milestone in church history coming on the heels of Vatican II, yet maintaining the traditional view that every act of sexual intercourse must be in accordance with the natural law, interpreted to mean that married love should respect rather than counter the natural cycles and rhythms present in human bodies, created by God.

HUMANAE VITAE

ENCYCLICAL OF POPE PAUL VI
ON THE REGULATION OF BIRTH

JULY 25, 1968

To His Venerable Brothers the Patriarchs, Archbishops, Bishops and other Local Ordinaries in Peace and Communion with the Apostolic See, to the Clergy and Faithful of the Whole Catholic World, and to All Men of Good Will.

Honored Brothers and Dear Sons, Health and Apostolic Benediction.

The transmission of human life is a most serious role in which married people collaborate freely and responsibly with God the Creator. It has always been a source of great joy to them, even though it sometimes entails many difficulties and hardships.

The fulfillment of this duty has always posed problems to the conscience of married people, but the recent course of human society and the concomitant changes have provoked new questions. The Church cannot ignore these questions, for they concern matters intimately connected with the life and happiness of human beings.

I. Problem and Competency of the Magisterium

2. The changes that have taken place are of considerable importance and varied in nature. In the first place there is the rapid increase in population which has made many fear that world population is going to grow faster than available resources, with the consequence that many families and developing countries would be faced with greater hardships. This can easily induce public authorities to be tempted to take even harsher measures to avert this danger. There is also the fact that not only working and housing conditions but the greater demands made both in the economic and educational field pose a living situation in which it is frequently difficult these days to provide properly for a large family.

Also noteworthy is a new understanding of the dignity of woman and her place in society, of the value of conjugal love in marriage and the relationship of conjugal acts to this love.

But the most remarkable development of all is to be seen in man's stupendous progress in the domination and rational organization of the forces of nature to the point that he is endeavoring to extend this control over every aspect of his own life, over his body, over his mind and emotions, over his social life, and even over the laws that regulate the transmission of life.

...

Interpreting the Moral Law

4. This kind of question requires from the teaching authority of the Church a new and deeper reflection on the principles of the moral teaching on marriage, a teaching which is based on the natural law as illuminated and enriched by divine Revelation.

No member of the faithful could possibly deny that the Church is competent in her magisterium to interpret the natural moral law. It is in fact indisputable, as Our predecessors have many times declared, (1) that Jesus Christ, when He communicated His divine power to Peter and the other Apostles and sent them to teach all nations His commandments, (2) constituted them as the authentic guardians and interpreters of the whole moral law, not only, that is, of the law of the Gospel but also of the natural law. For the natural law, too, declares the will of God, and its faithful observance is necessary for men's eternal salvation. (3)

In carrying out this mandate, the Church has always issued appropriate documents on the nature of marriage, the correct use of conjugal rights, and the duties of spouses. These documents have been more copious in recent times. (4)

Special Studies

5. The consciousness of the same responsibility induced Us to confirm and expand the commission set up by Our predecessor Pope John XXIII, of happy memory, in March, 1963. This commission included married couples as well as many experts in the various fields pertinent to these questions. Its task was to examine views and opinions concerning married life, and especially on the correct regulation of births; and it was also to provide the teaching authority of the Church with such evidence as would enable it to give an apt reply in this matter, which not only the faithful but also the rest of the world were waiting for.(5)

When the evidence of the experts had been received, as well as the opinions and advice of a considerable number of Our brethren in the episcopate, some of whom sent their views spontaneously, while others were requested by Us to do so, We were in a position to weigh with more precision all the aspects of this complex subject. Hence we are deeply grateful to all those concerned.

The Magisterium's Reply

6. However, the conclusions arrived at by the commission could not be considered by Us as definitive and absolutely certain, dispensing Us from the duty of examining personally this serious question. This was all the more necessary because, within the commission itself, there was not complete agreement concerning the moral norms to be proposed, and especially because certain approaches and criteria for a solution to this question had emerged which were at variance with the moral doctrine on marriage constantly taught by the magisterium of the Church.

Consequently, now that We have sifted carefully the evidence sent to Us and intently studied the whole matter, as well as prayed constantly to God, We, by virtue of the mandate entrusted to Us by Christ, intend to give Our reply to this series of grave questions.

II. Doctrinal Principles

7. The question of human procreation, like every other question which touches human life, involves more than the limited aspects specific to such disciplines as biology, psychology, demography or sociology. It is the whole man and the whole mission to which he is called that must be considered: both its natural, earthly aspects and its supernatural, eternal aspects. And since in the attempt to justify artificial methods of birth control many appeal to the demands of married love or of responsible parenthood, these two important realities of married life must be accurately defined and analyzed. This is what We mean to do, with special reference to what the Second Vatican Council taught with the highest authority in its Pastoral Constitution on the Church in the World of Today.

God's Loving Design

8. Married love particularly reveals its true nature and nobility when we realize that it takes its origin from God, who "is love," (6) the Father "from whom every family in heaven and on earth is named." (7)

Marriage, then, is far from being the effect of chance or the result of the blind evolution of natural forces. It is in reality the wise and provident institution of God the Creator, whose purpose was to effect in man His loving design. As a consequence, husband and wife, through that mutual gift of themselves, which is specific and exclusive to them alone, develop that union of two persons in which they perfect one another, cooperating with God in the generation and rearing of new lives.

The marriage of those who have been baptized is, in addition, invested with the dignity of a sacramental sign of grace, for it represents the union of Christ and His Church.

Married Love

9. In the light of these facts the characteristic features and exigencies of married love are clearly indicated, and it is of the highest importance to evaluate them exactly.

This love is above all fully human, a compound of sense and spirit. It is not, then, merely a question of natural instinct or emotional drive. It is also, and above all, an act of the free will, whose trust is such that it is meant not only to survive the joys and sorrows of daily life, but also to grow, so that husband and wife become in a way one heart and one soul, and together attain their human fulfillment.

It is a love which is total—that very special form of personal friendship in which husband and wife generously share everything, allowing no unreasonable exceptions and not thinking solely of their own convenience. Whoever really loves his partner loves not only for what he receives, but loves that partner for the partner's own sake, content to be able to enrich the other with the gift of himself.

Married love is also faithful and exclusive of all other, and this until death. This is how husband and wife understood it on the day on which, fully aware of what they were doing, they freely vowed themselves to one another in marriage. Though this fidelity of husband and wife sometimes presents difficulties, no one has the right to assert that it is impossible; it is, on the contrary, always honorable and meritorious.

The example of countless married couples proves not only that fidelity is in accord with the nature of marriage, but also that it is the source of profound and enduring happiness.

Finally, this love is fecund. It is not confined wholly to the loving interchange of husband and wife; it also contrives to go beyond this to bring new life into being. "Marriage and conjugal love are by their nature ordained toward the procreation and education of children. Children are really the supreme gift of marriage and contribute in the highest degree to their parents' welfare." (8)

Responsible Parenthood

10. Married love, therefore, requires of husband and wife the full awareness of their obligations in the matter of responsible parenthood, which today, rightly enough, is much insisted upon, but which at the same time should be rightly understood. Thus, we do well to consider responsible parenthood in the light of its varied legitimate and interrelated aspects.

With regard to the biological processes, responsible parenthood means an awareness of, and respect for, their proper functions. In the procreative faculty the human mind discerns biological laws that apply to the human person. (9)

With regard to man's innate drives and emotions, responsible parenthood means that man's reason and will must exert control over them.

With regard to physical, economic, psychological and social conditions, responsible parenthood is exercised by those who prudently and generously decide to have more children, and by those who, for serious reasons and with due respect to moral precepts, decide not to have additional children for either a certain or an indefinite period of time.

Responsible parenthood, as we use the term here, has one further essential aspect of paramount importance. It concerns the objective moral order which was established by God, and of which a right conscience is the true interpreter. In a word, the exercise of responsible parenthood requires that husband and wife, keeping a right order of priorities, recognize their own duties toward God, themselves, their families and human society.

From this it follows that they are not free to act as they choose in the service of transmitting life, as if it were wholly up to them to decide what is the right course to follow. On the contrary, they are bound to ensure that what they do corresponds to the will of God the Creator. The very nature of marriage and its use makes His will clear, while the constant teaching of the Church spells it out. (10)

Observing the Natural Law

11. The sexual activity, in which husband and wife are intimately and chastely united with one another, through which human life is transmitted, is, as the recent Council recalled, "noble and worthy.'" (11) It does not, moreover, cease to be legitimate even when, for reasons independent of their will, it is foreseen to be infertile. For its natural adaptation to the expression and strengthening of the union of husband and wife is not thereby suppressed. The fact is, as experience shows, that new life is not the result of each and every act of sexual intercourse. God has wisely ordered laws of nature and the incidence of fertility in such a way that successive births are already naturally spaced through the inherent operation of these laws. The Church, nevertheless, in urging men to the observance of the precepts of the natural law, which it interprets by its constant doctrine, teaches that each and every marital act must of necessity retain its intrinsic relationship to the procreation of human life. (12)

NOTES

LATIN TEXT: Acta Apostolicae Sedis, 60 (1968), 481–503.
ENGLISH TRANSLATION: The Pope Speaks, 13 (Fall. 1969), 329–46.

REFERENCES

(1) See Pius IX, encyc. letter Oui pluribus: Pii IX P.M. Acta, 1, pp. 9–10; St. Pius X encyc. letter Singulari quadam: AAS 4 (1912), 658; Pius XI, encyc.letter Casti connubii: AAS 22 (1930), 579–581; Pius XII, address Magnificate Dominum to the episcopate of the Catholic World: AAS 46 (1954), 671–672; John XXIII, encyc. letter Mater et Magistra: AAS 53 (1961), 457.
(2) See Mt 28. 18–19.
(3) See Mt 7. 21.
(4) See Council of Trent Roman Catechism, Part II, ch. 8; Leo XIII, encyc.letter Arcanum: Acta Leonis XIII, 2 (1880), 26–29; Pius XI, encyc.letter Divini illius Magistri: AAS 22 (1930), 58–61; encyc. letter Casti connubii: AAS 22 (1930), 545–546; Pius XII, Address to Italian Medico-Biological Union of St. Luke: Discorsi e radiomessaggi di Pio XII, VI, 191–192; to Italian Association of Catholic Midwives: AAS 43 (1951), 835–854; to the association known as the Family Campaign, and other family associations: AAS 43 (1951), 857–859; to 7th congress of International Society of Hematology: AAS 50 (1958), 734–735 [TPS VI, 394–395]; John XXIII, encyc.letter Mater et Magistra: AAS 53 (1961), 446–447 [TPS VII, 330–331]; Second Vatican Council, Pastoral Constitution on the Church in the World of Today, nos. 47–52: AAS 58 (1966), 1067–1074 [TPS XI, 289–295]; Code of Canon Law, canons 1067, 1068 §1, canon 1076, §§1–2.
(5) See Paul VI, Address to Sacred College of Cardinals: AAS 56 (1964), 588 [TPS IX, 355–356]; to Commission for the Study of Problems of Population, Family and Birth: AAS 57 (1965), 388 [TPS X, 225]; to National Congress of the Italian Society of Obstetrics and Gynecology: AAS 58 (1966), 1168 [TPS XI, 401–403].
(6) See 1 Jn 4. 8.
(7) Eph 3. 15.
(8) Second Vatican Council, Pastoral Constitution on the Church in the World of Today, no. 50: AAS 58 (1966), 1070–1072 [TPS XI, 292–293].
(9) See St. Thomas, Summa Theologiae, I–II, q. 94, art. 2.
(10) See Second Vatican Council, Pastoral Constitution on the Church in the World of Today, nos. 50–5 1: AAS 58 (1 966) 1070–1073 [TPS XI, 292–293].
(11) See ibid., no. 49: AAS 58 (1966), 1070 [TPS XI, 291–292].
(12) See Pius XI. encyc. letter Casti connubi: AAS 22 (1930), 560; Pius XII, Address to Midwives: AAS 43 (1951), 843.

— 8 —
Pontifical Academy of Life, Final Communique on the "Ethics of Biomedical Research for a Christian Vision" Feb. 26, 2003. Sections 1–4, 9, Manifesto.

Copyright – Libreria Edificie Vaticana and The Pontifical Academy of Life, 2003

This 2003 statement, issued by a new Pontifical Academy created by Pope John Paul II, includes mention of a number of key concepts in the Catholic Church's campaign to influence the burgeoning field of bioethics, such as each person's 'inalienable dignity' and the substantial, physical integrity of human persons. Such reasoning leads to the definition of early embryos as persons. This position is a challenge to stem cell research, which involves using and sometimes destroying human embryos.

PONTIFICAL ACADEMY FOR LIFE

CONCLUDING COMMUNIQUÉ ON THE
"ETHICS OF BIOMEDICAL RESEARCH.
FOR A CHRISTIAN VISION"

24–26 February 2003

1. The Ninth General Assembly of the Pontifical Academy for Life took place at the Vatican from 24–26 February. This year it was dedicated to a crucial theme that has a strong social impact, "Ethics of Biomedical Research. For a Christian Vision."

It is evident that, especially in the recent decades, biomedicine has developed in an extraordinary way, helped by the enormous progress in technology and computer science that have vastly extended the possibilities for experimentation on living beings and, especially on the human being. There have been tremendous breakthroughs, for example, in the fields of genetics, molecular biology, as well as in transplants and the neurological sciences.

Today more than ever, among the factors that contributed to this development, certainly biomedical research has been instrumental in the progress of knowledge in this sector of medicine, as the Holy Father himself recently pointed out: "It is a recognized fact that improvements in the medical treatment of disease primarily depend on progress in research" (John Paul II, Address to participants in the Ninth General Assembly of the Pontifical Academy for Life, 24 February 2003, n. 2; ORE, 5 March 2003, p. 4).

2. In the present setting, every new discovery in biomedicine seems destined to produce a "cascade" effect, opening up many new prospects and possibilities for the diagnosis and treatment of numerous pathologies that are still incurable.

Obviously, the acquisition of a growing technical possibility of intervention on human beings, on other living beings and on the environment, and the attainment of ever more decisive and permanent effects, obviously demands that scientists and society as a whole assume an ever greater responsibility in proportion to the power of intervention. It follows that the experimental sciences, and biomedicine itself, as "instruments" in human hands, are not complete in themselves, but must be directed to defined ends and put in dialogue with the world of values.

3. The primary agent of this continuous process of "ethical orientation" is, unmistakeably, the human person. Indivisible unity of body and soul, the human being is characterized by his capacity to choose in freedom and responsibility the goal of his own actions and the means to achieve it. His burning desire to seek the truth, that belongs to his nature and his specific vocation, finds an indispensible help in the Truth itself, God, who comes to meet the needs of the human being and reveals to him his Face through creation, and more directly, through Revelation. Thus God favours and supports the efforts of human reason, and enables the human being to recognize so many "seeds of truth" present in reality, and finally, to enter into communion with the Truth itself which He is.

In principle, therefore, there are no ethical limits to the knowledge of the truth, that is, there are no "barriers" beyond which the human person is forbidden to apply his cognitive energy: the Holy Father has wisely defined the human being as "the one who seeks the truth" (Fides et ratio, n. 28); but, on the other hand, precise ethical limits are set out for the manner the human being in search of the truth should act, since "what is technically possible is not for that very reason morally admissible" (Congregation for the Doctrine of the Faith, Donum Vitae, n. 4). It is therefore the ethical dimension of the human person, which he applies concretely though the judgements of his moral conscience, that connotes the existential goodness of his life.

4. In the commitment to research and to recognize the objective truth in every creature, a particularly important role falls to scientists in the area of biomedicine, who are called to work for the well-being and health of human beings, the ultimate aim of every research activity in this field must be the integral good of man. The means it uses, must fully respect every person's inalienable dignity as a person, his right to life and and his substantial physical integrity.

Against any false accusation or misunderstanding, let us repeat in communion with the Pope, John Paul II, that: "The Church respects and supports scientific research when it has a genuinely human orientation, avoiding any form of instrumentalization or destruction of the human being and keeping itself free from the slavery of political and economic interests" (Address to participants in the Ninth General Assembly of the Pontifical Academy for Life, 24 February 2003, n. 4; ORE, 5 March 2003, p. 4).

In this perspective, one must express the greatest possible gratitude to the thousands of doctors and researchers of the whole world who, generously and with great professionality, dedicate their energies every day to the service of the suffering and the treatment of pathologies. Further, the Pope recalled that: "all, believers and non–believers, acknowledge and express sincere support for these efforts in biomedical science that are not only designed to familiarize us with the marvels of the human body, but also to encourage worthy standards of health and life for the peoples of our planet" (ibid., n. 2) [. . .]

9. Special attention must also be paid to the treatment of human subjects who undergo research who are especially "vulnerable" because of their state of life, as the example of human embryos clearly illustrates. Because of the delicate stage of their development, possible experimentation on them in the light of current technological advances would involve a very high—and therefore ethically unacceptable—risk of causing them irreversible damage and even death.

The attitude some adopt concerning the legitimacy of sacrificing the (physical and genetic) integrity of human beings at the embryonic stage in order to destroy them, if necessary, in order to benefit other human individuals is likewise totally unacceptable. It is never morally licit to do evil intentionally in order to achieve ends that are good in themselves.

Moreover, it should be borne in mind that, although the human individual at the embryonic stage deserves the full respect that is due to every human person, human embryos are certainly not subjects who can give their personal consent to experimentation that exposes them to grave risks without the benefit of any directly therapeutic effect for themselves. Therefore, any experimentation on the human embryo that does not have the goal of obtaining direct benefits for his/her own health, cannot be considered morally licit [. . .]

Proposal of an Ethical Commitment for Researchers in the Biomedical Field

Introductory note

The following "manifesto" is published as an appendix to the Final Communiqué of the Ninth General Assembly of the Pontifical Academy for Life. It is a concrete result of the assembly's deliberations, whose theme this year was: "Ethics of Biomedical Research: For a Christian Vision," offered as an open proposal to be freely supported.

The invitation for a personal adherence is addressed to all researchers and those involved in research in the biomedical field and also to researchers in bioethics.

Those who wish to adhere to this "manifesto," which means that they embrace the principles it contains, to should communicate to the Academy by one of the following ways:

- by e—mail to: pav@acdlife.va
- by fax to: +39 06 69882014
- by post to: Pontificia Accademia per la Vita, Via della Conciliazione, 3, 00193 Rome, Italy.

Whatever the chosen modality, it is obligatory to include one's personal details (name, surname, address, telephone, fax, email), profession and place of employment, academic degrees and other qualifications.

Premise

The scientific developments of recent decades have brought about important cultural and social transformations, modifying in a qualitative way many aspects of human life. Indeed, the advance of scientific progress in many sectors has given rise to great hopes of concrete improvements for the life and future of the human person. However, in certain sectors of scientific research problems and/or doubts of an ethical and religious nature have arisen; they have demonstrated unequivocally the real need for constant dialogue/integration between the experimental sciences and the broader human sciences and philosophy in terms of operating in a more ample perspective so that the acquisition of greater knowledge may effectively serve the true good of the human person.

Human life and human nature appear to be realities too complex to be exhaustively evaluated from a single perspective; a multidisciplinary approach therefore appears indispensable for a better understanding of the human being in his integrity and contribute to a meaningful growth of a science that would truly be for the human being.

Moreover, such an interdisciplinary dialogue, by re—focusing attention on the centrality of the human person, would make the scientists more aware of the ethical implications of their work, and, conversely, would incite those involved in philosophical and theological anthropology to assume toward the scientists a mission of dialogue, collaboration and practical support, with the mutual intention of developing cognitive and applied tools for the service of the human community.

In this perspective, the reference to human values, and finally, to an anthropological and ethical vision, is an indispensable premise for a correct scientific research, that recognises the person's responsibility to himself and to others.

In fact, without reference to ethics, science and technology can be used either to kill or to save human lives, to manipulate or to promote, to destroy or to build. It is therefore necessary that, through responsible management, research be addressed toward the true common good, a good that transcends any merely private interest, going beyond the geographical and cultural boundaries of nations and keeping one's vision directed toward the good of future generations.

For science to be really placed at the service of the human being, it is necessary that it goes "beyond matter," intuiting in the corporeal dimension of the individual the expression of a greater spiritual good. Scientists should understand the human body as the tangible dimension of a unitary personal reality, which is at the same time corporeal and spiritual. The spiritual soul of the human being, although not in itself tangible, it is always the root of his existential and tangible reality, of his relationship with the rest of the world, and consequently, of his specific and inalienable value.

Only such a vision can make scientific research effectively respectful of the human person, considered in his complex corporeal–spiritual unity, every time he/she becomes the object of investigation, with particular reference to those events that constitute the beginning and the end of the individual human life.

For this reason, emerges a strong need to offer to young researchers formative programmes that put the accent not only on the scientific preparation, but also on the acquisition of the fundamental notions of anthropology and ethics. The expression of such programmes could, then, crystallize in the elaboration of a true and proper Deontological Code for researchers, to which each researcher could safely refer in his work, and which, at the same time, would represent a sign of hope and commitment for a truly "humanized" medicine in the new millennium.

A first indication of the way to take, might concern the manner in which the researchers should behave and the norms they should observe in order to direct their research towards the objective just recalled above. It is our desire to propose such ethical indications, to which we firmly adhere, to all others who are involved in the world of biomedical research; somehow, they delineate the principal features of the researcher's "moral personality."

Commitment
- I commit myself to adhere to a methodology of research characterized by scientific rigour and a high quality of the information that is furnished.
- I will not take part in research projects in which I could be subject of a conflict of interests, from the personal, professional or economic point of view.
- I recognise that science and technology must be at the service of the human person, fully respecting his dignity and rights.
- I recognise and respect all researches and their applications which are based on the principle of "moral goodness" and referring to the correct vision of the corporeal and spiritual dimensions of the human being.

- I recognise that every human being, from the first moment of his existence (process of fertilization) up to the moment of his natural death, is to be guaranteed the full and unconditional respect due to every human person by virtue of his peculiar dignity.

- I recognise, because of my duty to safeguard human life and health, the usefulness and the obligation of a serious and responsible experimentation on animals, carried out according to determined ethical guidelines, before applying new diagnostic and therapeutic methodologies to human beings. I also recognise that the passage from the experiments with animals to the clinical experimental stage (on man) should take place only when the evidences resulting from the experiments with animals sufficiently demonstrate the harmlessness or the acceptability of the possible harms and risks that such experiments might involve.

- I recognise the legitimacy of clinical experiments on the human being, but only under precise conditions, including, in the first place, the safeguarding of the life and physical integrity of human beings who are involved. Then, there is the need that the experiments be always preceded by proper, correct and complete information regarding the significance and developments of the same experiments. I will treat each person who submits to an experiment as a free and responsible subject and never as a mere means to achieve other ends. I will never let a person be involved in an experiment unless he/she has given his/her free and informed consent.

Glossary

Accident: In Aristotelian philosophy, accidents are the perceptible qualities of an object such as its color, texture, size, shape, etc.

Alchemy: From Arabic *al-kimia*, it refers to both a pre-scientific form of the investigation of nature and a spiritual or philosophical discipline. Alchemy combined elements of metallurgy, medicine, and astrology with elements of what would become physics and chemistry.

Angular unconformity: A separation between two contiguous rock strata of different ages, indicating that sediment deposition was not continuous. The rocks above an unconformity are younger than the rocks beneath (unless the sequence has been overturned). An unconformity represents time during which no sediments were deposited. The phenomenon of angular unconformities was discovered by James Hutton, who found examples at Jedburgh in 1787 and at Siccar Point in 1788.

Anthropic principle: The idea that, based on a statistically improbable consilience of cosmological constants such as the ratio of the mass of the proton to the electron and the velocity of light, the universe is intentionally designed for human beings to emerge within the evolutionary process. While this argument and evidence for it present profound difficulties for atheists, theologians are sometimes reluctant to draw on it as proof for God's creation of the universe because of the possibility that the consilience of factors might not be so improbable given future research into the early life of the universe shortly after the "Big Bang."

Apologist: One who defends a position against attack. In the Early Church era, the apologists were theologians aiming to show that Christianity was consistent with reason.

Atomism: In natural philosophy, Greco-Roman theory of matter (Epicurus and Lucretius) that all objects in the universe are composed of very small, indestructible building blocks (Greek *atomos* means, "that which cannot be cut into smaller pieces.") It was revived in early modernity by Pierre Gassendi and René Descartes.

Catastrophism: Dominant view before Lyell that the earth was formed by catastrophic events. Georges Cuvier made no reference to religion in his writings, and assumed periodic catastrophes from the stratigraphic record. By contrast, flood geologists read the earth's geological history in light of the biblical food story.

College of Cardinals: The body of church officials (usually bishops) whose function is to elect the pope and to advise him about ecclesiastical matters.

Deep time: The concept of "geological time" orders of magnitude greater than the approximately 6,000 years of "scriptural time." In the West, it was first recognized in the eighteenth century by James Hutton, Charles Lyell, and others.

Deism: A form of belief in God that developed in Enlightenment France, England, and the United States. In contrast to theism, in deism God does not interfere with individual human lives or the laws of the universe.

Enlightenment: Eighteenth-century philosophical movement, also known as the "Age of Reason," which advocated reason as the primary basis of authority. It influenced political thinking that led to the American and French Revolutions (the "Bill of Rights" and the "Declaration of the Rights of Man and the Citizen"); deism was one religious response.

Evolution: Theory that species can change over time. Darwinian variant: descent from common ancestry with modification. Lamarckian variant: change through the transmission of acquired characteristics.

Galenic: Galen was a brilliant second-century Greek physician whose medical and physiological ideas dominated the scholastic tradition until they were disproven by Andreas Vesalius (1514–1564) and William Harvey (1578–1657).

Heliocentric or heliostatic theory: the idea, first broached by Aristarchus in the fourth century BCE that the earth and other planets or "wandering stars" revolve about a stationary sun. The theory was submerged by the Aristotelian-Ptolemaic model of a geocentric or earth-centered cosmos.

Historical critical method: Scholarly approach to biblical study. Lower criticism tries to establish the original text by comparing extant manuscripts; higher criticism examines the authorship, social context, and literary history of a text.

Index: List of Prohibited Books *(Index Librorum Prohibitorum)* promulgated by Pope Paul IV in 1564 and regularly updated, which included works regarded as potentially harmful to the faithful for reasons of moral or doctrinal error. The Index was abolished in 1966 under Pope Paul VI.

Inquisition: Ecclesiastical tribunal for inquiry into and suppression of heresy; some were temporary of shorter or longer duration (e.g., the Spanish Inquisition). In 1542 Pope Paul III established the Sacred Congregation of the Holy Office to maintain and defend the integrity of the faith.

Intelligent Design: The theory advocated by several American scientists and philosophers (including Catholic biochemist Michael Behe) that the "irreducibly complex" structures of biochemical and biological systems yield the hypothesis of their design by an external nonmaterial agent (usually understood to refer to God) instead of change through the standard evolutionary mechanism of natural selection with random mutation.

Lemaître, Georges: Belgian, Jesuit priest and astronomer (1894–1966) who developed a model of the universe that would conform to Einstein's theories of relativity and which was later termed "the Big Bang." He later felt obliged to oppose Pope Pius XII's attempt, in a 1951 speech, to identify the moment of divine creation of the universe with the Big Bang.

Modernism: The object of Pope Pius X's attack in his encyclical letter *Pascendi Dominici Gregis* (1907). In the view of its opponents, modernism stood for (1) Enlightenment secularism, (2) a rationalist approach to biblical criticism, (3) modern philosophical systems, and (4) the view that the Catholic Church and its doctrines and practices have evolved over time and can change radically.

Monasticism: A religious practice in which one renounces worldly pursuits in order to devote one's life fully to spiritual work (Greek: *monachos*—a solitary person). Benedict of Nursia found his monastery of Monte Cassino in 529, and the movement flowered in the Middle Ages.

Natural Law: The theory that certain principles of moral behavior are "inscribed" on the human heart and which direct human acting inasmuch as human beings act on the basis of an informed conscience. In this way, human beings can identify the general framework for action, what one "ought" to do, in how nature is characterized. Catholic moral doctrine teaches that morality is applied on the basis of both natural law and the Christian ethics contained in revelation (scripture and church teaching).

Nebular hypothesis: Cosmological theory that the solar system was formed from a gaseous cloud (in Latin *nebula* means "clouds"). First proposed in 1734 by Emanuel Swedenborg; Immanuel Kant proposed in 1755 that nebulae slowly rotate, gradually collapsing and flattening due to gravity and eventually forming stars and planets. A mathematically more precise model was proposed in 1796 by Pierre-Simon Laplace.

Neo-Darwinian synthesis: Integration of Darwin's theory of the evolution of species through natural selection, Mendel's theory of genetics as the basis for biological inheritance, random genetic mutation as the source of variation, and mathematical population genetics.

Neoplatonism: A school of religious and mystical philosophy that took shape in the third century CE, based on the teachings of Plato and earlier Platonists. Neoplatonism influenced Augustine, Bonaventure, and some Renaissance thinkers.

Original Sin: The historic Christian doctrine that sin is a condition into which human beings have a tendency to fall from Adam onwards. This weakness is cited as either because of a general weakness of will or a biologically inherited property of human beings from generation to generation.

Physico-theology: Seventeenth- and eighteenth-century development of medieval "natural theology" (as opposed to "revealed theology") illustrating the existence and attributes of God from nature alone. Popular in Britain as a way of harmonizing theology with the new science.

Pius IX, Pope: (May 13, 1792 to February 7, 1878), born Giovanni Maria Mastai-Ferretti, he reigned from 1846–1878, making him the longest-reigning Pope since the Apostle St. Peter. Originally moderate and sympathetic to democratic and modernizing reforms in Italy and in the Church, Pius became increasingly conservative after he was deposed as the temporal ruler of the Papal States and drew up the *Syllabus of Errors*. He formally defined the dogma of the Immaculate Conception and organized the First Vatican Council, which defined the dogma of papal infallibility.

Pius XII, Pope: Succeeded Pope Pius XI in 1939 and reigned until 1958. A keen observer of scientific developments, he is best known for acknowledging the success of evolutionary theory in his 1950 encyclical *Humani Generis*.

Scholasticism: From Latin *scholasticus*—"that [which] belongs to the school." A method of learning taught by medieval universities (ca. 1100–1500) emphasizing dialectical reasoning as a way to reconcile Aristotelian philosophy with medieval Christian theology. It developed two genres of literature: (1) Posing of *quæstiones* or questions (e.g., "Is it permissible to kill for self-preservation?); the student could refer to any number of sources to find the pros and cons of the question. (2) A *summa* or complete summary of what it was possible to deduce on the subject.

Stratigraphy: The branch of geology that studies the layering of rocks, or stratification.

Substance: In Aristotelian philosophy What something is in itself, the substratum underlying all qualities.

Syllabus of Errors: A document issued by the Vatican under Pope Pius IX in 1864. It was and remains controversial for its condemnation of concepts such as freedom of religion and the separation of church and state.

Teilhard de Chardin, Pierre: French, Jesuit priest and paleontologist (1881–1955). Teilhard, as is known, developed an interpretation of cosmic and biological evolution which stresses purposeful, divinely warranted emergence of complex living systems in the universe, whose total existence is harmonic, not competitive as with standard interpretations of Darwinism. He integrates a spiritual vision of nature with insights concerning the special character of human rationality and the unique role of Jesus Christ in cosmic evolution. His thought was repressed and his books forbidden to be published until after his death at which time his fame soared.

Theodicy: Philosophical or theological explanation to justify the existence of a benevolent, omnipotent, and omniscient God in the face of the reality of suffering.

Thomism: A philosophical school of thought named after scholastic theologian and philosopher Thomas Aquinas (1224–1274), it was revived in the nineteenth century by Pope Leo XIII after which it became widely known as neo-Thomism. It is a philosophical system, which interprets Aristotle's philosophy of substance and form, potency and act in a way that stresses the order of the world as a purposeful, created reality. While it suffered a decline in seminaries and Catholic universities after Vatican II, it has made a modest comeback in the thought of some contemporary Catholic thinkers.

Transubstantiation: (in Latin, *transsubstantiatio*) in the theology of the Roman Catholic Church, this refers to the *change of the substance* of bread and wine into the body and blood of Christ occurring in the Eucharist, while the accidents (color, taste, texture) remain the same. Depends upon Aristotelian philosophical categories.

Trent, Council of: Nineteenth council recognized by Catholics; met in three sessions (1545–1563) in the city of Trent in northern Italy as a response to the theological and ecclesiological challenges of the Protestant Reformation. Specified Catholic doctrines on salvation, the sacraments, and the Biblical canon; standardized the Mass throughout the church (the "Tridentine Mass").

Trivium and Quadrivium: In medieval scholasticism these formed the foundation of the undergraduate arts curriculum: the "three ways" (Logic, grammar, rhetoric or dialectic) and the "four ways" (arithmetic, geometry, astronomy, and music).

Ultramontane: Literally, "living beyond the mountains," usually south of the alps. Northern Europeans had developed a tendency to regard the papacy as a foreign power, especially when the Pope interfered in temporal matters by favoring some ruler or country over another. In the nineteenth century ultramontanism implied conservatism and papal control.

Uniformitarianism: Coined by William Whewell in 1832, it refers to the view of James Hutton and Charles Lyell that that the natural processes operating in the past are the same as those that can be observed operating in the present, e.g., glaciers carving valleys. Summarized in the slogan, "the present is the key to the past."

Vatican I, Council: Convened by Pope Pius IX in 1968; first session held in Saint Peter's Basilica on December 8, 1869; 800 church leaders attended. In three sessions, there was discussion and approval of only two constitutions: the *Dogmatic Constitution On The Catholic Faith,* and *Pastor Aeternus,* the *First Dogmatic Constitution on the Church of Christ,* dealing with the primacy and papal infallibility.

Vatican II, Council: A gathering of the world's Catholic bishops that took place over four sessions between 1962 and 1965 (twenty-first ecumenical council). Presided over by Pope John XXIII until his death in 1963 and thereafter by Pope Paul VI, Vatican II ushered in a series of changes and reforms to church governance and to formulations in numerous doctrinal areas (liturgical renewal; theology of the Church and its relation to the modern world, scriptural studies; ecumenical dialogue toward reconciliation with other Christian churches).

Bibliography

Abbott, Walter M., S.J., ed. *The Documents of Vatican II.* New York: Herder and Herder, 1966.
Abraham, Carolyn. "New stem cell technique may help solve political, ethical debate." *The Globe and Mail* (Toronto), August 24, 2006. Available at http://www.theglobeandmail.com/servlet/story/RTGAM.20060823.wstemcell0823/BNStory/Science/home (Last accessed on January 5, 2007).
Acta Apolostica Sedis (Official Documents of The Holy See, Vatican City) 54 (1962).
Alszeghy, Zoltán, S.J. "Development in the doctrinal formulation of the Church concerning the theory of evolution." In *The Evolving World and Theology*, ed. Johannes Metz, *Concilium*, Vol. 26. New York: Paulist Press, 1967, citing *Collectio Lacensis*, 5, 292.
Appleby, R. Scott. "Exposing Darwin's 'hidden agenda': Roman Catholic responses to evolution, 1875–1925." In *Disseminating Darwinism: The Role of Place, Race, Religion, and Gender*, ed. Ronald L. Numbers and John Stenhouse. Cambridge: Cambridge University Press, 1999, 173–207.
Aquinas, Thomas. *Summa Contra Gentiles*, trans. Joseph Rickaby, S.J. London: Burns and Oates, 1905.
Aristotle. *Meteorologica* (Loeb Classical Library No. 397), trans. H. D. P. Lee. Cambridge, MA: Harvard University Press, 1975.
———. *The Metaphysics*, trans. Hugh Lawson-Tancred. Penguin Classics, new ed. London: Penguin Books, 1999.
Artigas, Mariano, and Shea, William R. *Galileo in Rome: The Rise and Fall of a Troublesome Genius.* Oxford: Oxford University Press, 2003.
Artigas, Mariano, Thomas F. Glick, and Rafael A. Martinez. *Negotiating Darwin: The Vatican Confronts Evolution, 1877–1902.* Baltimore: The Johns Hopkins University Press, 2006.
Ashworth, William. "Catholicism and early modern science." In *God and Nature: Historical Essays on the Encounter between Christianity and Science*, ed. David

Bibliography

C. Lindberg and Ronald L. Numbers. Berkeley: University of California Press, 1986, 136–166.

Astruc, Jean. *Conjectures sur les mémoires originauz dont il paroit que Moyse s'est servi pour composer le livre de la Génèse* ("Conjectures on the original documents that Moses appears to have used in composing the Book of Genesis") Brussels, 1753.

Augustine. *The City of God*. Translated by Marcus Dods. New York: Modern Library, 1993.

Augustine. *Enchiridion, on Faith, Hope and Love*, in *On Christian Belief*, ed. Boniface Ramsey, trans. Matthew O'Connell. Hyde Park, NY: New City Press, 2005.

Augustine of Hippo. *De genesi ad litteram (The Literal Meaning of Genesis)*, trans. J. H. Taylor, S.J., Book I, Chapter 19. New York: Newman Press, 1982.

———. *Confessions*, trans. Henry Chadwick. Oxford: Oxford University Press, 1998.

Baldini, Ugo. "The Academy of Mathematics of the Collegio Romano from 1553 to 1612." In *Jesuit Science and the Republic of Letters*, ed. Mordechai Feingold. Cambridge, MA: The MIT Press, 2003, 47–69.

Barbour, Ian. *Religion and Science: Historical and Contemporary Issues*. San Francisco: HarperSanFrancisco, 1997.

Barr, Stephen. Review of Stanley Jaki's *Bible and Science*. *First Things* 71 (March, 1997), 46–48.

Behe, Michael. *Darwin's Black Box*. New York: The Free Press, 1998.

Benedict of Nursia. *Life and Miracles of St. Benedict (Book II, Dialogues)*, trans. Odo John Zimmerman, O.S.B., and Benedict R. Avery, O.S.B. Westport, CT: Greenwood Press, 1980.

———. *Rule of Saint Benedict*, trans. John Baptist Hasbrouck. Kalamazoo, MI: Cistercian Publications, 1983.

Berry, Thomas. *The Dream of the Earth*. San Francisco: Sierra Club Books, 1988.

Blackwell, Richard J. *Bellarmine, and the Bible: Including a Translation of Foscarini's Letter on the Motion of the Earth*. Notre Dame, IN: University of Notre Dame Press, 1991.

Boscovich, Ruggero. *Theoria philosophiae naturalis reducta ad unicam legem virium in natura existentium* (Theory of natural philosophy reduced to a single law of force), Venice or Vienna 1758.

Bossy, John. *The English Catholic Community, 1570–1850*. London: Darton, Longman and Todd, 1975.

Bouwsma, William J. *The Waning of the Renaissance, 1550–1640*. New Haven, CT: Yale University Press, 2000.

Bowler, Peter J. *The Mendelian Revolution: The Emergence of Hereditarian Concepts in Modern Science and Society*. Baltimore: Johns Hopkins University Press, 1989.

———. *Evolution: The History of an Idea*, 3rd ed. Berkeley: University of California Press, 2003.

Bracken, Joseph. "*Quaestio Disputata*: Response to Elizabeth Johnson's 'Does God Play Dice?'" *Theological Studies* 57 (1996), 720–730.

Brennan, Martin S. *The Science of the Bible*. St. Louis: B. Herder, 1898.

Brennan, Robert E. "Troubadour of truth." In *Essays in Thomism*, ed. R.E. Brennan. New York: Sheed and Ward, 1942, 1–25.

Brooke, John H. *Science and Religion: Some Historical Perspectives*. Cambridge: Cambridge University Press, 1991.

———. "Revisiting Darwin on order and design." In *Design and Disorder: Perspectives from Science and Theology*, ed. Niels Gregersen and Ulf Gorman. London: T & T Clark, 2002, 31–52.

Brooke, John, and Cantor, Geoffrey. *Reconstructing Nature: The Engagement of Science and Religion*. Edinburgh: T&T Clark, 1998.

Browne, Janet. *The Secular Ark: Studies in the History of Biogeography*. New Haven: Yale University Press, 1983.

Browne, Janet E. *Charles Darwin: Voyaging*. Princeton University Press, 1996.

Buckley, Michael J. *At the Origins of Modern Atheism*. New Haven, CT: Yale University Press, 1987.

Burrow, J. W. *The Crisis of Reason: European Thought, 1848–1914*. New Haven, CT: Yale University Press, 2000.

Byers, David ed. *Religion, Science and the Search for Wisdom: Proceedings of a Conference on Religion and Science*. Washington: United States Catholic Conference, 1987.

Campbell, Kenneth L. *The Intellectual Struggle of the English Papists in the Seventeenth Century: The Catholic Dilemma*. Lewiston, NY: Edwin Mellen Press, 1986.

The Catechism of the Roman Catholic Church (CCC) [English translation]. New York: Doubleday, 1994.

Clifford, Anne. "Catholicism and Ian Barbour on theology and science." In *Fifty Years in Science and Religion: Ian G. Barbour and His Legacy*, ed. Robert J. Russell. Aldershot, UK: Ashgate, 2004, 287–300.

Cohen, H. Floris. "The onset of the scientific revolution: Three near-simultaneous transformations." In *The Science of Nature in the Seventeenth Century: Patterns of Change in Early Modern Natural Philosophy*, ed. Peter Anstey and John A. Schuster Dordrecht, the Netherlands: Springer-Verlag, 2005, 9–34.

Cohn, Norman. *Noah's Flood: The Genesis Story in Western Thought*. New Haven, CT: Yale University Press, 1996.

Colish, Martha. *Medieval Foundations of the Western Intellectual Tradition, 400–1400*. New Haven, CT: Yale University Press, 1997.

Copernicus, Nicolaus. *De revolutionibus orbium coelestium libri sex* (1543).

———. *On the Revolutions*, translation and commentary by Edward Rosen. Baltimore, MD: Johns Hopkins University Press, 1992.

Costello, William T., S.J. *The Scholastic Curriculum at Early Seventeenth-Century Cambridge*. Cambridge, MA: Harvard University Press, 1958.

Crombie, A. C. *Robert Grosseteste and the Origins of Experimental Science*. Oxford: The Clarendon Press, 1953.

Cutler, Alan. *The Seashell on the Mountaintop: A Story of Science, Sainthood and the Humble Genius Who Discovered a New History of the Earth*. London: Penguin/Dutton, 2003.

Dahm, J. J. "Science and apologetics in the early Boyle Lectures." *Church History*, 39 (1970), 172–186.

Danielson, Dennis R. "The great Copernican cliché." *American Journal of Physics*, 69(10) (2001): 1029–1035.

Darwin, Charles. *The Origin of Species*, 6th ed. London: Oxford University Press, 1872, reprinted 1956.

Deane-Drummond, Celia. *Genetics and Christian Ethics*. Cambridge: Cambridge University Press, 2005.

Deason, B. Gary. "Reformation theology and the mechanistic conception of nature." In *God and Nature: Historical Essays on the Encounter between Christianity and Science*, ed. David C. Lindberg and Ronald L. Numbers. Berkeley: University of California Press, 1986, 167–191.

Deltete, Robert J. "Pierre Duhem's *Physique de Croyant*." Paper delivered at the 2004 annual meeting of the Society for the History of the Philosophy of Science (HOPOS), University of San Francisco, California, June 2004.

De Lubac, Henri. *Teilhard Explained*. New York: Paulist Press, 1968.

Desmond, Adrian, and James Moore. *Darwin: The Life of a Tormented Evolutionist*. New York: W. W. Norton, 1994.

Dick, Steven J. *Plurality of Worlds: The Origins of the Extraterrestrial Life Debate from Democritus to Kant*. Cambridge: Cambridge University Press, 1982.

Dillenberger, John. *Protestant Thought and Natural Science*. Notre Dame, IN: University of Notre Dame Press, 1988.

Dinis, Giovanni. "Giovanni Battista Riccioli and the science of his time." In *Jesuit Science and the Republic of Letters*, ed. Mordechai Feingold. Cambridge, MA: The MIT Press, 2003, 195–224.

Dizionario Interdisciplinare di Scienza e Fede. Rome: Urbaniana University Press – Città Nuova, 2002. On-line dynamic database at http://www.disf.org/en/default.asp.

Dogmatic Constitution on the Catholic Faith (April 1870), IV 9.

Dooley, Brendan. "The *Storia Letteraria D'Italia* and the rehabilitation of Jesuit science." In *Jesuit Science and the Republic of Letters*, ed. Mordechai Feingold. Cambridge, MA: The MIT Press, 2003, 433–473.

Dorlodot, Canon Henry de. *Darwinism and Catholic Thought*, Vol. 10, trans. Ernest Messenter. London: Burns, Oates and Washbourne, 1922.

Dreyer, John L. *Tycho Brahe: A Picture of Scientific Life and Work in the Sixteenth Century*. Edinburgh: A. and C. Black, 1890; rpt. Whitefish, MT: Kessinger Publishing, 2004.

Duhem, Pierre. *Origines de la Statique*. Paris: Librairie Scientifique A Hermann, 1906.

———. "Physique de Croyant." In *La Théorie Physique, son Objet et sa Structure*. Paris: Éditions Marcel Rivière, 1906; second edition 1914.

du Noüy, Lecomte. *Human Destiny*. London: Longmans, Green and Co., 1947.

Dupree, A. Hunter. "Christianity and the scientific community in the age of Darwin." In *God and Nature: Historical Essays on the Encounter between Christianity and Science*, ed. David C. Lindberg and Ronald Numbers. Berkeley: University of California Press, 1986, 351–368.

Edwards, Denis. "Original sin and saving grace in evolutionary context." In *Evolutionary and Molecular Biology: Scientific Perspectives on Divine Action*, ed. Robert John Russell, William Stoeger, S.J., and Francisco Ayala. CTNS/Vatican City: Vatican Observatory Foundation, 1998, 377–392.

Eisenstein, Elizabeth L. *The Printing Press as an Agent of Change*. Cambridge: Cambridge University Press, 1979.

Elder, Gregory P. *Chronic Vigour: Darwin, Anglicans, Catholics, and the Development of a Doctrine of Providential Evolution*. New York: University Press of America, 1996.

Erasmus, Desiderius. *The Colloquies of Erasmus*, Vol. I, trans. N. Bailey. London: BiblioBazaar, 2007.

Eternal Word Television Network. Available at: http://www.ewtn.com/library/PAPALDOC/JP961022.HTM (Last accessed on May 14, 2006).

Faherty, William B. "John Hagen: Eminent European astronomer sojourns in Wisconsin." *Wisconsin Magazine of History*, (1941): 178–186.

Fantoli, Annibale. *Galileo, for Copernicanism and for the Church*, 3rd ed., trans George V. Coyne. Vatican City: Vatican Observatory Publications; Notre Dame, IN: University of Notre Dame Press, 2003.

Feingold, Mordechai. "Jesuits: Savants." In *Jesuit Science and the Republic of Letters*, ed. Mordechai Feingold. Cambridge, MA: The MIT Press, 2003, 1–45.

Feldhay, Rivka. *Galileo and the Church: Political Inquisition or Critical Dialogue?* Cambridge, MA: Cambridge University Press, 1995.

Ferngren, Gary B., ed. *Science and Religion: An Historical Introduction*. Baltimore: The Johns Hopkins University Press, 2002.

Findlen, Paula. "Courting nature." In *Cultures of Natural History*, ed. N. Jardine, J. A. Secord, and E. C. Spary. Cambridge: Cambridge University Press, 1996, 57–74.

Flannery, Austin, O.P. ed. *The Basic Sixteen Documents. Vatican Council II, Constitutions, Decrees Declarations*. Northport, NY: Costello Publications, 1996.

Frend, W. H. C. *The Donatist Church: A Movement of Protest in Roman North Africa*. Oxford: Clarendon Press, 1952.

Fuller, Reginald C. *Alexander Geddes: A Pioneer of Biblical Criticism, 1737–1802*. Sheffield, England: Almond Press, 1984.

Galilei, Galileo. *Discoveries and Opinions of Galileo*, trans. Stillman Drake. Garden City, NY: Doubleday, 1957.

———. *Sidereus Nuncius, or The Sidereal Messenger*, trans. Albert van Helden. Chicago: University of Chicago Press, 1989.

Garrigan, Owen. *Man's Intervention in Nature*. New York: Hawthorn Books, 1967.

Gaukroger, Stephen. *The Emergence of a Scientific Culture: Science and the Shaping of Modernity, 1210–1685*. Oxford: Clarendon Press; New York: Oxford University Press, 2006.

Genuth, Sara Schechner. *Comets, Popular Culture, and the Birth of Modern Cosmology*. Princeton, NJ: Princeton University Press, 1997.

Gingerich, Owen. *An Annotated Census of Copernicus' De revolutionibus* (Nuremberg, 1543 and Basel, 1566). Leiden: Brill Academic Publishers, 2002.

———. *The Book Nobody Read: Chasing the Revolutions of Nicolaus Copernicus*. New York: Walker & Company, 2004.

———. "The Censorship of Copernicus' De Revolutionibus." Florence: *Annali dell'Instituto e Museo di Storia della Scienza di Firenze* 7, 1981:45–61.

———. "Galileo: Hero or heretic?" Paper presented at the *"Faith, Science, and the Future" Symposium*, Concordia College, Moorhead, MN, 1998.

Gleason, Elizabeth G. *Gasparo Contarini: Venice, Rome, and Reform*. Berkeley: University of California Press, 1993.

———. *Reform Thought in Sixteenth-Century Italy*. Chico, CA: Scholars Press, 1981.

Glick, Thomas F, ed. "Religion and Darwinism: Varieties of Catholic reaction." In *The Comparative Reception of Darwinism*. Austin: The University of Texas, 1972, 403–436.

Gonzalez, Justo L. *The Story of Christianity*, Vol. 2. New York: Harper & Row, 1984.

Goodwin, C. W. "The Mosaic cosmogony," in Hedge, ed. *Essays and Reviews*, London: J.W. Parker and Son, 1860.

Grafton, Anthony, and Nancy Siraisi, eds. *Natural Particulars: Nature and the Disciplines in Renaissance Europe*. Cambridge, MA: The MIT Press, 1999.

Grant, Edward. "Celestial perfection from the Middle Ages to the late seventeenth century." In *Rethinking the Scientific Revolution*, ed. Margaret Osler. Cambridge: Cambridge University Press, 2000, 137–162.

———. *The Foundations of Modern Science in the Middle Ages: Their Religious, Intellectual and Institutional Contexts*. Cambridge: Cambridge University Press, 1996.

———. *God and Reason in the Middle Ages*. Cambridge: Cambridge University Press, 2001.

———. "Science and theology in the Middle Ages." In *God and Nature: Historical Essays on the Encounter between Christianity and Science*, ed. David C. Lindberg and Ronald L. Numbers. Berkeley: University of California Press, 1986, 49–75.

———. *A Sourcebook in Medieval Science*. Cambridge, MA: Harvard University Press, 1974.

Hahn, Roger. *Pierre Simon Laplace, 1749–1827: A Determined Scientist*. Cambridge, MA: Harvard University Press, 2005.

Hankins, James. "The study of the *Timaeus* in early Renaissance Italy." In *Natural Particulars: Nature and the Disciplines in Renaissance Europe*, ed. Anthony Grafton and Nancy Siraisi. Cambridge, MA: The MIT Press, 1999, 77–119.

Harrison, Peter. "Scaling the ladder of being: Theology and early theories of evolution." In *Religion, Reason, and Nature in Early Modern Europe*, ed. Robert Crocker. Dordrecht, the Netherlands: Kluwer Academic Publishers, 2001, 199–224.

Hastings, Adrian ed. *Modern Catholicism: Vatican II and After*. Oxford: Oxford University Press, 1990.

Hauber, W. A. "Evolution and Catholic thought." *American Ecclesiastical Review*, 106 (1942):161–177.

Haude, Sigrun. "Anabaptism." In *The Reformation World*, ed. Andrew Pettegree. London: Routledge, 2000a, 237–256.

———. *In the Shadow of "Savage Wolves": Anabaptist Münster and the German Reformation during the 1530s*. Leiden: Brill Academic Publishers, 2000b.

Haught, John F. *Science and Religion: From Conflict to Conversation*. Mahwah, NJ: Paulist Press, 1995.

———. *God After Darwin.* Boulder, CO: Westview Press, 2000.
———. *Deeper Than Darwin: The Prospect for Religion in an Age of Evolution.* Boulder, CO: Westview Press, 2003.
———. "What if theologians took evolution seriously?" *New Theology Review*, 18(4) (2005): 10–20.
———. *Is Nature Enough? Meaning and Truth in the Age of Science.* Cambridge: Cambridge University Press, 2006.
Hedge, Frederick H. *Essays and Reviews by Eminent English Churchmen* (1874).
Heilbron, John L. *The Sun in the Church: Cathedrals as Solar Observatories.* Cambridge, MA: Harvard University Press, 1999.
Henig, Robin Marantz. *The Monk in the Garden: How Gregor Mendel and His Pea Plants Solved the Mystery of Inheritance.* Boston: Houghton Mifflin, 2000.
Hess, Hamilton. *The Early Development of Canon Law and the Council of Serdica.* Oxford: Oxford University Press, 2002.
Hess, Peter M. J. "Copernicus, Nicolaus." In *The Encyclopedia of Religion*, Vol. 3. Detroit: Macmillan Reference, 2004, 1976–1979.
———. "God's two books: Special revelation and natural science in the Christian West." In *Bridging Science and Religion*, ed. Ted Peters. Minneapolis, MN: Fortress Press, 2003, 123–140.
———. "Natural history in relation to religion from Aristotle to Darwin." In *Science and Religion: An Historical Introduction*, ed. Gary B. Ferngren. Baltimore: The Johns Hopkins University Press, 2002, 195–207.
Hiebert, Erwin R. "Modern physics and Christian faith." In *God and Nature: Historical Essays on the Encounter between Christianity and Science*, ed. David C. Lindberg and Ronald L. Numbers. Berkeley: University of California Press, 1986, 424–447.
Howell, Kenneth J. *God's two books: Copernican Cosmology and Biblical Interpretation in Early Modern Science.* Notre Dame, IN: University of Notre Dame Press, 2002.
Hsia, R. Po-chia. *The World of Catholic Renewal, 1540–1770.* Cambridge: Cambridge University Press, 2005.
Hurlbut, William. *Presidential Commission on Bioethics Session 6: Seeking Morally Unproblematic Sources of Human Embryonic Stem Cells.* Transcript, December 3, 2004. Available at http://www.bioethics.gov/transcripts/dec04/session6.html (Last accessed on January 25, 2007).
Hutton, James. *System of the Earth*, 1785. *Theory of the Earth*, 1788. *Observations on granite*, 1794. Together with *Playfair's Biography of Hutton*. Intro. by Victor A. Eyles. Darien, CN: Hafner Publishing Co., 1970.
Ignatius of Antioch, "Epistle to the Smyrneans." *The Ante-Nicene Fathers*, vol. 1, ed. Alexander Roberts and James Donaldson. Grand Rapids, MI: William B. Eerdmans, 1979, 86–92.
Iserloh, Erwin, Joseph Glazik, and Hubert Jedin. *Reformation and Counter Reformation*, trans. Anselm Biggs and Peter W. Becker. *History of the Church*, Vol. 5. New York: The Seabury Press, 1980.
Jaki, Stanley. *The Road of Science and the Ways to God.* Chicago: University of Chicago Press, 1978.

———. *Scientist and Catholic: An Essay on Pierre Duhem.* Fort Royal, VA: Christendom Press, 1991.
———. *Bible and Science.* Fort Royal, VA: Christendom Press, 2004.
Janacek, Bruce. "Catholic natural philosophy: Alchemy and the revivification of Sir Kenelm Digby." In *Rethinking the Scientific Revolution*, ed. Margaret J. Osler. Cambridge: Cambridge University Press, 2000.
Jedin, Hubert. *Katholische Reformation oder Gegenreformation?* [*Catholic Reformation or Counter-Reformation?*]. Luzern: Josef Stocker, 1946.
Jesuits in Science, Newsletter (1997) Available at http://www.jesuitsinscience.org/Newsletter97/papal.htm (Last accessed on May 14, 2006).
Johnson, Elizabeth. "Does God play dice? Divine providence and chance." *Theological Studies*, 57 (1996): 3–18.
Jones, D. Gareth. *Teilhard de Chardin: An Analysis and Assessment.* Grand Rapids, MI: Eerdmans, 1970.
Kaiser, Christopher. *Creational Theology and the History of Physical Science: The Creationist Tradition from Basil to Bohr.* Leiden: Brill, 1997.
Kargon, R. H. *Atomism in England from Hariot to Newton.* Oxford: Clarendon, 1966.
King, Thomas, and James F. Salmon, eds. *Teilhard and the Unity of Knowledge, The Georgetown University Centennial Symposium.* New York: Paulist Press, 1983.
Kingsley, Charles. *Prose Idylls, New and Old.* London: Macmillan, 1889.
Knabenbauer, J. "Stimmen aus Maria-Laach." XIII (1877), 74; Cited in *Catholic Encyclopedia* (1914), http://www.newadvent.org/cathen/04470a.htm.
Kragh, Helga. "Big bang cosmology." In *Cosmology: Historical, Literary, Philosophical, Religious, and Scientific Perspectives*, ed. Norriss Hetherington. New York: Garland Publishing, 1993, 371–389.
Kuhn, Thomas S. *Copernican Revolution: Planetary Astronomy in the Development of Western Thought.* Cambridge, MA: Harvard University Press, 1977.
Küng, Hans. *On Being a Christian.* London: Collins, 1976.
———. *Does God Exist? An Answer for Today*, trans. Edward Quinn. New York: Vintage Books, 1981.
Ladrière, Jean. "Faith and cosmology." In *Language and Belief*, trans. Garrett Bardin. Notre Dame, IN: University of Notre Dame Press, 1972, 149–186.
———. "Meaning and truth in theology." *Catholic Theological Society of America Proceedings*, 42 (1987): 1–15.
———. "The role of philosophy in the science-theology dialogue." In *The Interplay between Scientific and Theological Worldviews I*, ed. Niels H. Gregersen, Ulf Görman, and Christoph Wassermann. Geneva: Labor et Fides, 1999.
Langford, Jerome J. *Galileo, Science and the Church*, 3rd. ed. Ann Arbor: The University of Michigan Press, 1992.
Laplace, Pierre Simon. *Mécanique céleste*, 4 vols. Paris: Duprat, 1799–1805.
Larson, Edward. *Evolution: The Remarkable History of a Scientific Theory.* New York: Modern Library, 2004.
Lattis, James M. *Between Copernicus and Galileo: Christoph Clavius and the Collapse of Ptolemaic Cosmology.* Chicago: University of Chicago Press, 1994.
LeClerq, Jean, OSB. *The Love of Learning and the Desire for God.* New York: Fordham University Press, 1961.

Leff, Gordon. "The *trivium* and the three philosophies," in *A History of the University in Europe*, ed. Hilde De Ridder-Symoens. Cambridge: Cambridge University Press, 1992, 307–336.

Lindberg, David C. *The Beginnings of Western Science: The European Scientific Tradition in Philosophical, Religious, and Institutional Context, 600 B.C. to A.D. 1450*. Chicago: University of Chicago Press, 1992.

———. "Galileo, the Church, and the cosmos." In *When Science and Christianity Meet*, ed. David C. Lindberg and Ronald L. Numbers. Chicago: University of Chicago Press, 2003a, 33–60.

———. "The Medieval Church encounters the classical tradition: Saint Augustine, Roger Bacon, and the handmaiden metaphor." In *When Science and Christianity Meet*, ed. David C. Lindberg and Ronald L. Numbers. Chicago: University of Chicago Press, 2003b, 7–32.

———. "Science and the Early Church." In *God and Nature: Historical Essays on the Encounter between Christianity and Science*, ed. David C. Lindberg and Ronald L. Numbers. Berkeley: University of California Press, 1986, 19–48.

Lindberg, David C., and Ronald L. Numbers, eds. *God and Nature: Historical Essays on the Encounter between Christianity and Science*. Berkeley: University of California Press, 1986.

———. *When Science and Christianity Meet*. Chicago: University of Chicago Press, 2003.

Logan, Alastair H. B. *The Gnostics: Identifying an Early Christian Cult*. London: T&T Clark, 2006.

Lonergan, Bernard. *Insight: A Study of Human Understanding*, 5th ed. F.E. Crowe and R. M. Doran, eds. In *Collected Works of Bernard Lonergan*, Vol. 3. Toronto: University of Toronto Press, 1992.

Lyell, Charles. *Principles of Geology*, 4 vols. 3rd ed., May 1834; 2 vols. 12th ed., 1875 (published posthumously).

MacDonnell, Joseph, S.J. *Jesuit Geometers: A Study of Fifty-Six Prominent Jesuit Geometers during the First Two Centuries of Jesuit History*. Vatican City: The Vatican Observatory, 1989.

Martineau, James. "Science, nescience, and faith." *National Review* 15 (1862): 394–419; in *Essays, Reviews, and Addresses*, Vol. 3. London: Longmans, Green and Co., 1891.

Martini-Bettòlo, G. B. *Discourses of the Popes from Pius XI to John Paul II to the Pontifical Academy of Sciences 1936–1986*. Vatican: Pontificia Academia Scientiarum, 1986a.

———. *Historical Aspects of the Pontifical Academy of Sciences*. Vatican: Pontificia Academia Scientiarum, 1986b.

May, William. *Catholic Bioethics and the Gift of Human Life*. Huntingdon, IN: Our Sunday Visitor, 2000.

McCalla, Arthur. *The Creationist Debate: The Encounter between the Bible and the Historical World*. London: T & T Clark, 2006.

McClymond, Michael J. "Jesus." In *The Rivers of Paradise: Moses, Buddha, Confucius, Jesus, and Muhammad as Religious Founders*, ed. David Noel Freedman and Michael J. McClymond. Grand Rapids, MI: Eerdmans, 2001, 356–375.

McMullin, Ernan, ed. *Evolution and Creation*. Notre Dame, IN: University of Notre Dame Press, 1985.

———. "How should cosmology relate to theology?" In *The Sciences and Theology in the Twentieth Century*, ed. Arthur Peacocke. Notre Dame, IN: University of Notre Dame Press, 1981, 17–57.

———. "Natural science and belief in a creator." In *Physics, Philosophy and Theology: A Common Quest for Understanding*, ed. Robert J. Russell, William Stoeger, S.J., and George Coyne. Vatican City: CTNS/Vatican Observatory, 1988, 49–79.

———. Review of John Polkinghorne's *Belief in God in an Age of Science*. *Commonweal* 125(17) (October 9, 1998).

Meier, John P. *A Marginal Jew: Rethinking the Historical Jesus, Vol. 1: The Roots of the Problem and the Person*. New York: Doubleday, 1991.

Midbon, Mark. "A day without yesterday: Georges Lemaitre & the Big Bang." *Commonweal* (March 24, 2000).

Miller, Kenneth. *Finding Darwin's God*. New York: HarperCollins, 1999.

Mivart, St. George Jackson. *On the Genesis of Species*, 2nd ed. London: Macmillan, 1871.

Moore, Aubrey L. *Science and the Faith: Essays on Apologetic Subjects*. London: Kegan Paul, Trench, Trübner, 1893.

Moore, James R. "Geologists and interpreters of Genesis in the nineteenth century." In *God and Nature: Historical Essays on the Encounter between Christianity and Science*, ed. David C. Lindberg and Ronald L. Numbers. Berkeley: University of California Press, 1986, 322–350.

Muckerman, H. "Biogenesis and abiogenesis." *Catholic Encyclopedia*, 1907.

Mullett, Michael A. *The Catholic Reformation*. London: Routledge, 1999.

Murdoch, John E. "From social into intellectual factors: An aspect of the unitary character of late medieval learning." In *The Cultural Context of Medieval Learning*, ed. John E. Murdoch and Edith Dudley Sylla. Dordrecht, Holland: D. Reidel Publishing Company, 1975, 271–348.

Murphy, Nancey. "Introduction." In *The Neurosciences and the Person: Scientific Perspectives on Divine Action*, ed. Robert Russell, Nancey Murphy, Theo Meyering and Michael Arbib. Vatican City: CTNS/Vatican Observatory Foundation, 1999, i–xxxv.

Murray, Alexander. "Nature and man in the Middle Ages." In *The Concept of Nature: The Herbert Spencer Lectures*, ed. John Torrance. Oxford: The Clarendon Press, 1992, 25–62.

Navarro, Victor. "Tradition and scientific change in early modern Spain: The role of the Jesuits." In *Jesuit Science and the Republic of Letters*, ed. Mordechai Feingold. Cambridge, MA: The MIT Press, 2003, 331–367.

Newman, John Henry. "Christianity and physical science." In *The Idea of a University Defined and Illustrated*. New York: Longmans, Green, and Co., 1907, Chapter 7.

Nicholson, Marjorie H. *The Breaking of the Circle: Studies in the Effect of the "New Science" upon Seventeenth-Century Poetry*. New York: Columbia University Press, 1960.

Noonan, John T. *Contraception: A History of Its Treatment by the Catholic Theologians and Canonists.* New York: New American Library, 1967.

North, John. "The *quadrivium*," in *A History of the University in Europe*, ed. Hilde De Ridder-Symoens. Cambridge: Cambridge University Press, 1992, 337–359.

Oakes, Edward. "Final causality: A response." *Theological Studies*, 53(3) (September 1992): 534–544.

———. Interview, Zenit News Agency "Interview with Fr. Edward Oakes, S.J." July 28, 2005 (www.zenit.org) (Last accessed on August 15, 2005).

O'Connor, J.J., and E. F. Robertson. "Ruggero Giuseppe Boscovich." The MacTutor History of Mathematics Archive, http://www-groups.dcs.st-and.ac.uk/~history/Biographies/Boscovich.html

O'Leary, Don. *Roman Catholicism and Modern Science: A History.* New York: Continuum International Publishing Group, 2006.

Olin, John C. *Catholic Reform: From Cardinal Ximenes to the Council of Trent, 1495–1563.* New York: Fordham University Press, 1990.

O'Malley, John W. *The First Jesuits.* Cambridge, MA: Harvard University Press, 1993.

Oresme, Nicole. *Le Livre du ciel et du monde*, 1377. ed. Albert D. Menut and Alexander J. Denomy, trans. and introd. Albert D. Menut. Madison, WI: University of Wisconsin Press, 1968.

Osler, Margaret. "Baptizing Epicurean atomism: Pierre Gassendi on the immortality of the soul. In *Rethinking the Scientific Revolution*, ed. Margaret Osler. Cambridge: Cambridge University Press, 2000, 163–183.

———. "The canonical imperative: Rethinking the Scientific Revolution." In *Rethinking the Scientific Revolution*, ed. Margaret Osler. Cambridge: Cambridge University Press, 2000, 3–22.

Ozment, Steven. *The Age of Reform, 1250–1550: An Intellectual and Religious History of Late Medieval and Reformation Europe.* New Haven, CT: Yale University Press, 1980.

Pambrun, James R. "Theology, modern science and the mediations of meaning: A reflection on the contribution of Jean Ladrière." *Laval Théologique et Philosophique*, 57(3) (October 2001): 469–493.

Pascal, Blaise. *Pensées*, ed. and trans. Roger Ariew. Indianapolis, IN: Hackett Publishing. Co., 2005.

Paul, Harry W. *The Edge of Contingency: French Catholic Reaction to Scientific Change from Darwin to Duhem.* Gainesville: University Presses of Florida, 1979.

Pelikan, Jaroslav. *The Christian Tradition: A History of the Development of Doctrine, Vol. 1: The Emergence of the Catholic Tradition.* Chicago: University of Chicago Press, 1971.

———. *The Christian Tradition: A History of the Development of Doctrine, Vol. 4: Reformation of Church and Dogma (1300–1700).* Chicago: University of Chicago Press, 1984.

Plato, *Phaedo*, trans. R. S. Bluck. London: Routledge & Kegan Paul, 1955.

Pohle, J. "Angelo Secchi." In *Catholic Encyclopedia.* New York: Robert Appleton Company, 1912a. On the Internet at http://www.newadvent.org/cathen/13669a.htm

———. *God the Author of Nature and the Supernatural*. London: B. Herder Book Company, 1912b.

Polanyi, Michael. *Personal Knowledge: Towards a Post-Critical Philosophy*. New York: Harper Torchbooks, 1964.

Pontifical Academy of Life, Profile. See http://www.vatican.va/roman_curia/pontifical_academies/acdlife/ (Last accessed on April 24, 2007).

Pontifical Academy of Life. *Reflections on Cloning*. Vatican City: Libreria Editrice Vaticana, 1997.

Pope John Paul II. "Truth Cannot Contradict Truth" Address to the Pontifical Academy of the Sciences, October 22, 1996. Available at: http://www.newadvent.org/library/docs_jp02tc.htm

———. "Faith can never conflict with reason," Address to the Pontifical Academy of Sciences, October 31, 1992. *L'Osservatore Romano* no. 1264 (November 4, 1992).

———. *Fides et Ratio* 1998.

———. General Audience: January 26, 2000. Zenit News Agency. See http://www.zenit.org

Pope Leo XIII. *Providentissimus Deus:* Encyclical of Pope Leo XIII on the Study of Holy Scripture. Vatican City: Holy See, 1893.

Pope Pius IX. *Dogmatic Constitution on the Catholic Faith*. Vatican City, 1870a.

———. *Dogmatic Constitution on the Church*. Vatican City, 1870b.

———. *Syllabus Errorum (The Syllabus of Errors)*. Vatican City, 1864. http://www.papalencyclicals.net/Pius09/p9syll.htm

Pope Pius X. *Pascendi Dominici Gregis*, Encyclical on the Doctrine of the Modernists. September 8, 1907.

Pope Pius XII. *Divino Afflante Spiritu,* Vatican City, 1943.

———. *Humani Generis*. Vatican City, 1951.

Pope, Stephen J. *The Evolution of Altruism and the Ordering of Love*. Washington D.C.: Georgetown University Press, 1994.

Popkin, Richard H. *The History of Skepticism from Savonarola to Bayle*. 4th edition. New York and Oxford University Press, 2003.

Randles, W. G. L. *The Unmaking of the Medieval Christian Cosmos, 1500–1760: From Solid Heavens to Boundless Æther*. Aldershot, England: Ashgate Publishers, 1999.

Rashdall, Hastings. *The Universities of Europe in the Middle Ages*, 3 vols., revised by F. M. Powicke, and A. B. Emden. Oxford: Clarendon Press, 1987.

Rausch, Thomas. *Reconciling Faith and Reason: Apologists, Evangelists and Theologians in a Divided Church*. Minneapolis, MN: Liturgical Press, 2000.

Ray, John. *The Wisdom of God Manifested in the Works of Creation*, 1691. Reprint New York: Arno Press, 1977.

Repcheck, Jack. *The Man Who Found Time: James Hutton and the Discovery of the Earth's Antiquity*. Cambridge, MA: Perseus Publishing, 2003.

Reuters, "Scientists grow adult stem cells from nose." (March 24, 2005).

Rhonheimer, Martin. "The Truth about Condoms." *The Tablet* (July 10, 2004): 10–11.

Roberts, Michael B. "Genesis Chapter 1 and geological time from Hugo Grotius and Marin Mersenne to William Conybeare and Thomas Chalmers (1620–1825)."

In *Myth and Geology*, ed. L. Piccardi and W. B. Masse, Special Publication No. 273. Geological Society of London (March 15, 2007), 39–49.

Roger, Jacques. "The mechanistic conception of life." In *God and Nature: Historical Essays on the Encounter between Christianity and Science*, ed. David C. Lindberg and Ronald L. Numbers. Berkeley: University of California Press, 1986, 277–295.

Rosen, Edward. *Copernicus and the Scientific Revolution*. Malabar, FL: Krieger Press, 1984.

Rossi, Paolo. *The Dark Abyss of Time: The Theory of the Earth and the History of Nations from Hooke to Vico*, trans. Lydia G. Cochrane. Chicago: University of Chicago Press, 1984.

Rowland, Ingrid D. *The Ecstatic Journey: Athanasius Kircher in Baroque Rome*. Chicago: University of Chicago Library, 2000.

Rudwick, Martin J.S. "The shape and meaning of earth history." In *God and Nature: Historical Essays on the Encounter between Christianity and Science*, ed. David C. Lindberg and Ronald L. Numbers. Berkeley: University of California Press, 1986, 296–321.

———. *Scenes from Deep Time: Early Pictorial Representations of the Prehistoric World*. Chicago: University of Chicago Press, 1992.

———. *Bursting the Limits of Time: The Reconstruction of Geohistory in the Age of Revolution*. Chicago: University Of Chicago Press, 2007.

Scharper, Stephen B. "The ecological crisis." In *The Twentieth Century: A Theological Overview*, Gregory Baum, ed. Maryknoll, NY: Orbis Books, 1999, 219–227.

Schechner, Sara J. *Comets, Popular Culture, and the Birth of Modern Cosmology*. Princeton, NJ: Princeton University Press, 1997.

Schmitt, Charles B. *Aristotle and the Renaissance*. Cambridge, MA: Harvard University Press, 1983.

Schönborn, Cardinal Christoph. "Finding design in nature." *The New York Times* (July 7, 2005), A23.

Schroeder, H. J., ed. *The Canons and Decrees of the Council of Trent*. Rockford, IL: Tan Books, 1978.

Sergeant, John. *The Method to Science*. London: W. Redmayne, 1696.

———. *Transnatural Philosophy, or Metaphysics: Demonstrating the Essences and Operations of All Beings Whatever, Which Gives the Principles to all Other Sciences. And Showing the Perfect Conformity of Christian Faith to Right Reason, and the Unreasonableness of Atheists, Deists, Anti-trinitarians, and Other Sectaries* (London, 1700).

Shapiro, Barbara J. *Probability and Certainty in Seventeenth-Century England: A Study of the Relationships between Natural Science, Religion, History, Law, and Literature*. Princeton, NJ: Princeton University Press, 1983.

Shea, William R. "Galileo and the Church." In *God and Nature: Historical Essays on the Encounter between Christianity and Science*, ed. David C. Lindberg and Ronald L. Numbers. Berkeley: University of California Press, 1986, 114–135.

Sherrard, Philip. *The Greek East and the Latin West: A Study in the Christian Tradition*. Limni, Greece: D. Harvey, 1995.

Sobel, Dava. *Galileo's Daughter: A Historical Memoir of Science, Faith, and Love.* New York: Walker & Co., 1999.
Sorondo, Marcelo Sánchez. "Pontifical Academy of Sciences." In *Interdisciplinary Encyclopedia of Religion and Science*, Rome: Urbaniana University Press–Città Nuova, 2002, http://www.disf.org/en/Voci/93.asp.
Sowle Cahill, Lisa. "Bioethics." *Theological Studies*, 67 (2006): 120–142.
Spalding, John L. *Religion, Agnosticism, and Education.* Chicago: A. C. McClurg & Co., 1902.
Spencer, Herbert. *The Study of Sociology.* London: Williams and Norgate, 1874.
Spini, Giorgio. "The Rational of Galileo's Religiousness." In *Galileo Reappraised*, ed. Carlo L. Golino. Berkeley: University of California Press, 1966.
Stanford Encyclopedia of Philosophy, "Thomas Kuhn," http://plato.stanford.edu/entries/thomas-kuhn/ (Last accessed on September 16, 2006).
Stock, Brian. "Experience, praxis, work, and planning in Bernard of Clairvaux: Observations on the *Sermones in Cantica*." In *The Cultural Context of Medieval Learning*, ed. John E. Murdoch and Edith D. Sylla. Dordrecht, Holland: D. Reidel Publishing, 1975, 218–268.
Stoeger, William. "Contemporary cosmology and its implications for the science-religion dialogue." In *Physics, Philosophy and Theology: A Common Quest for Understanding*, ed. Robert J. Russell, William Stoeger, S.J., and George Coyne, eds. Vatican City: Vatican Observatory Foundation, 1988, 219–247.
———. "The immanent directionality of the evolutionary process, and its relationship to teleology." In *Evolutionary and Molecular Biology: Scientific Perspectives on Divine Action*, ed. Robert Russell, William Stoeger, and Francisco Ayala. Vatican City: Vatican Observatory Foundation, 1998, 163–190.
Synan, Edward A. "Introduction." In *Albertus Magnus and the Sciences: Commemorative Essays, 1980.* Toronto: Pontifical Institute of Medieval Studies, 1980.
Teilhard de Chardin, Pierre. *The Phenomenon of Man.* New York: Harper, 1975.
Temple, Frederick. *The Relations between Religion and Science.* London: Macmillan, 1884.
Tertullian. *The Prescription against Heretics*, trans. Peter Holmes, in *The Ante-Nicene Fathers*, Vol. 3, ed. Alexander Roberts and James Donaldson. New York: Charles Scribner's Sons, 1903. Grand Rapids, MI: William B. Eerdmans, 1978, 243–265.
Tilley, Maureen A. *The Bible in Christian North Africa: The Donatist World.* Minneapolis, MN: Fortress Press, 1997.
Trevor-Roper, H. R. "Nicholas Hill, the English atomist." In *Catholics, Anglicans and Puritans: Seventeenth-Century Essays.* Chicago: University of Chicago Press, 1987, 1–39.
University of California Museum of Palaeontology Web site, http://www.ucmp.berkeley.edu/history/cuvier.html.
Vatican Council I. *Dogmatic Constitution on the Catholic Faith*, April 1870.
Vatican Observatory Web site, http://clavius.as.arizona.edu/vo/R1024/History_p1.html.
Wallace, William. *Galileo, the Jesuits, and the Medieval Aristotle.* Hampshire, UK: Variorum, 1991.

---. "Science and religion in the Thomistic tradition (1)." *The Thomist*, 65 (2001): 441–463.
Wallace, W. A. *The Scientific Methodology of Theodoric of Freiberg. A Case Study of the Relationship between Science and Philosophy*, Studia Friburgensia, N.S. 26. Fribourg: The University Press, 1959.
Walsh, James J. *Catholic Churchmen in Science*. Philadelphia: American Ecclesiastical Review, 1909.
Walsh, Michael. "Pius XII." In *Modern Catholicism: Vatican II and After*, ed. Adrian Hastings. London: SPCK, 1991, 20–27.
Ware, Timothy. *The Orthodox Tradition*, rev. ed. London: Penguin Books, 1997.
Weisheipl, James A. "The life and works of St. Albert the Great." In *Albertus Magnus and the Sciences: Commemorative Essays, 1980*. Toronto: Pontifical Institute of Medieval Studies, 1980.
Westfall, Richard S. "The Scientific Revolution of the seventeenth century: The construction of a new world view." In *The Concept of Nature: The Herbert Spencer Lectures*, ed. John Torrance. Oxford: The Clarendon Press, 1992, 63–93.
Westman, Robert S. "The Copernicans and the churches." In *God and Nature: Historical Essays on the Encounter between Christianity and Science*, ed. David C. Lindberg and Ronald L. Numbers. Berkeley: University of California Press, 1986, 76–113.
White, Andrew Dickson. *A History of the Warfare of Science with Theology in Christendom*. New York: D. Appleton, 1897; reprint London: Bibliobazaar, 2007.
Whitehead, A. N. *Process and Reality*, corrected ed. New York: Free Press, 1978.
Windle, Bertram. *The Catholic Church and Its Relations with Science*. New York: MacMillan, 1927.
Wildiers, N. Max. *The Theologian and His Universe: Theology and Cosmology from the Middle Ages to the Present*. New York: Seabury Press, 1982.
Woods, Henry. *Augustine and Evolution, a Study in the Saint's De genesi ad litteram and De trinitate*. New York: The Universal Knowledge Foundation, 1924.
Yates, Frances A. *Giordano Bruno and the Hermetic Tradition*. London: Routledge and Kegan Paul, Ltd., 1964; rpt. Chicago: University of Chicago Press, 1991.
Zahm, John. *Evolution and Dogma*. Chicago: D. H. McBride and Co., 1896; reprint New York: Arno Press, 1978.

Index

Abortion, 158
Accademia de Lincei, 39–40, 47, 55, 64
Accommodation, principle of, 45, 70
Adam and Eve, Fall of, 14
Aeterni Patris, 100
Agassiz, Louis, 84
Albert the Great, 17
Alchemy, 55
Alighieri, Dante, 33, 59
Ambrose of Milan, 9
Americanism, 82, 85
Analogy, 126
Angular unconformity, 61–62
Anthony of Egypt, Saint, 13
Anthropic principle, 125, 129
Apologists, 7
Aquinas, Saint Thomas, 15, 17–18; and scriptural accommodation, 76; and teleology, 65
Aquinas, Thomas: in seminaries, 88, 121; and Thomism, 92–97
Archimedes, 32, 43
Aristarchus of Samos, 32
Aristotelianism: as empirical tool, 16; four causes, 16, 39; as reactionary, 40, 47; substance physics, of, 52–53; vital powers in, 67

Aristotle: *Physics*, 17; recovery of his thought, 15, 22–23; rejected by Spalding, 85; and science, 4–5; and spontaneous generation, 63; and teleology, 67; and Thomism, 94, 186
Artificial nutrition and hydration, 162–63
Astrology, 11, 55
Athanasius, 186–87
Atheism, 53, 58, 67; Mivart's opposition to, 76, 81
Atomic theory of matter, 46, 50, 53–55, 57, 67
Augustine of Hippo, 8–11; cited by Galileo in *Letter to Grand Duchess Christina*, 174; *De Genesi ad Litteram*, 63, 185; and original sin, 102, 168–69; rejection of astrology, 11
Augustinian Order, 26, 86
Averroes, 19

Babbage, Charles, 66
Bacon, Sir Francis, 39, 64
Bacon, Roger, 13, 16–17
Barbarian invasions, 12; of tenth century, 14
Barbour, Ian, 49, 58, 119; and Ernan McMullin, 134–35

Baronius, Cesare, 49
Bartholomeus Anglicus, 14
Behe, Michael, 130–32
Bellarmine, Robert Cardinal, 45
Benedict of Nursia, 13
Bernard of Clairvaux, 14
Berry, Thomas, 143–44
Bessell, Friedrich, 43
Bestiaries, 14–15
Biblical interpretation: authority over, 46–47; and "higher" criticism, 67–68; Pius IX and, 74
Big Bang cosmology/theory, 93, 104–6
Billings Method, The, 153
Bioethics, 158; and biomedical research, 205–11
Birth control, 150
Boethius of Dacia, 19
Bologna, University of, 15, 31
Bonaventure, Saint, 19
Boscovich, Ruggero, 57–58
Boyle Lectures, 65
Boyle, Robert, 51, 55
Bracken, Joseph, 145–47
Brahe, Tycho, 36, 38
Brennan, Martin S., and *The Science of the Bible*, 83
Bridgewater Treatises, 66
Bruno, Giordano, 29
Buckland, William, 69
Buffon, Comte de, 67
Byzantium, 6

Cahill, Lisa Sowle, 163
Calvin, Jean, 25
Cambridge University, 62, 65
Catastrophism, in geology, 69
Catholic: term first used, 1–2; church as a unifying force, 38, 41
Catholic Reformation, 28
Cesi, Federico, 39–40, 47–48, 64
Chalcedon, Council of, 7
Charles V, Emperor, 27–28
Chemistry, 55
Christianity, attitude to nature, 7
Christina, Grand Duchess, 173

Chronicles of the world, 70
Cistercian Order, 14
Civiltà Cattolica, La, 83
Clavius, Christopher, 30, 40
Cloning, 62, 161
College of Cardinals, 74
Collegio Romano, 50, 56–57
Cologne, Council of, 73
Cologne Declaration, The, 152
Comets, 31, 38, 45; Halley's, 57
Condemnations of 1277, 19–20, 38
Conflict model, 41, 45, 58
Consonance, 134–35
Constance, Council of, 28
Constantine, Emperor, 2, 6, 39
Contact model, 57
Contrast model, 58
Copernicanism, 40–41, 43, 45–46
Copernicus, Nikolaus, 25, 31–38, 62, 174, 176–77
Corpuscularian philosophy. *See* "Atomism"
Counter Reformation, 28, 40, 47, 59
Coyne, George, 124
Cranmer, Thomas, 25
Critical realism, 134
Cuvier, Georges, 69

Dalmace, Leroy, 80
Damascene, John, 65
Darwin, Charles, 62–63, 66, 70–71; never placed on *Index*, 83, 128; *On the Origin of Species*, 71, 74
Darwin, Erasmus, 70
Dawson, William, 81
Deane-Drummond, Celia, 167–70
Death, 166
"Deep time," 62; discovery of, 69–71
Deism, 57, 65, 70, 185
Dennett, Daniel, 131
Denza, Francesco, 56
De revolutionibus orbium coelestium libri sex, 35, 41, 45, 58
Derham, William, and *Physico-Theology, or, a Demonstration of the Being and Attributes of God*, 65

Descartes, René: and matter theory, 54, 67; vortex theory of, 54
Design argument, 65
Dialogue Concerning the Two Chief World Systems, 45, 48
Dialogue model, 73
Diet of Speyer, 27
Diet of Worms, 27
Diocletian persecution, 6
Discourse on Two New Sciences, 45
Divine Comedy, 33, 59
Divine Providence, 54, 184
Divino Afflante Spiritu, 76, 99
Dogmatic Constitution on the Catholic Faith, 74
Dogmatic Constitution on the Church, 74–75
Dominican Order, 17, 48
Donatists, 2
Dorlodot, Henri de, 89
Duhem, Pierre, 58–59; and Stanley Jaki, 142
Du Noüy, Lecomte, 131
Dürer, Albrecht, 64

Ecclesiastical corruption, 26
Ecology, 155–57; and Thomas Berry, 143–44
Eddington, Arthur, and Georges Lemaître, 104
Edict of Milan, 6
Edwards, Denis, 170
Eichhorn, Johann Gottfried, 68
Einstein, Albert, 57; and Georges Lemaître, 104
Elliptical planetary orbits, 38
Emblematic worldview, 50, 64
Emergentism, 166–67
Enlightenment, 57; and natural history, 66–71; Scottish, 65; secularism, 88
Epicurus, 53
Epicycles, 46
Erasmus, Desiderius, 39
Essays and Reviews, 73
Euthanasia, 162

Evil: problem of, 65; and animal suffering, 78
Evolution, theory of, 77, 79; Brennan's careful rejection of, 83; no Catholic conspiracy against, 83; nuance regarding Catholic response, 88; Zahm's contrast of Darwinism with, 81
EWTN television network, 122
Extinction, discovery of, 69

Final cause, 65, 67
Florence, 45, 48
Fossils, 64; incompleteness of noted by Brennan, 84; paucity of transitional fossils recognized by Zahm, 82
Foucault's pendulum, 43
Francis of Assisi, 183
Franciscan Order, 19
Frederick the Wise, Prince, 27
French Revolution, 74, 79
Frombork Cathedral, 31, 35
Füchsel, C. G., 70

Galilei, Galileo, 39, 41–49, 67; excerpt from *Letter to Grand Duchess Christina*, 173–77; Newman on, 73; and Pope John Paul II, 114
Gassendi, Pierre, 49, 54
Geddes, Alexander, 68
Genetics, Mendelian, 82, 167–68
Genesis, Book of, 68, 88–89
Gesner, Conrad, and *Historia Animalium*, 64
Gnosticism, 7–8
Gonzáles, Zeferino Cardinal, 78
Gould, Stephen Jay, 110
Gray, Asa, 84
Gregorian calendar reform, 30, 40
Grew, Nehemiah, *Cosmologia Sacra*, 64
Grosseteste, Robert, 16

H. M. S. *Beagle*, 62
Haeckel, Ernst, 83
Hagen, Johann, 57

Hauber, William, 89
Haught, John, 57–58, 137–38
Hebrew religion, 3–4
Hefner, Philip, 169
Heliocentric hypothesis, 31–35, 43, 45
Hell, Mivart's theology of, 78
Henry VIII, King, 27
Holy Cross Order, 80
Holy Spirit, 175
Hooke, Robert, 64
Hoyle, Fred, 105
Humanae Vitae, 123, 151
Humani Generis, 93, 100–103, 187–90
Humanism, 38–41, 68
Hume, David, and *Dialogues Concerning Natural Religion*, 65
Husserl, Edmund, 132
Hutton, James, 61–62, 69; *Theory of the Earth*, 62, 70
Huxley, Thomas, 76, 83

Ignatius of Antioch, 1
Ignatius of Loyola, 30
Index of Prohibited Books, 29–30, 45, 50, 55, 58, 78, 80, 82
Indulgences, 26
Infallibility, papal, 74
Inquisition (Holy Office), 29, 48, 50
Intelligent Design theory, 65, 128–29; and Michael Behe, 130–32
Isidore of Seville, 14
Islamic culture, 6

Jagiellonian University, 31
Jaki, Stanley, 141–43
James, William, 94
Jesuits, Order, 30; dissolution of, 58; and Galileo, 43, 48, 50; and the *Kulturkampf*, 79; in Science, 51, 122
Jesus of Nazareth, 3
Johannsen, Wilhelm, 88
John of Leiden, 29
Johnson, Elisabeth, 145–46

Kant, Immanuel, 58; *Critique of Teleological Judgment*, 67

Kelvin, Lord (William Thomson), 71, 84
Kepler, Johannes, 38, 46, 48
Kircher, Athanasius, 50
Knabenbauer, J., 79
Knox, John, 25
Kuhn, Thomas, 123, 136
Küng, Hans, 123–24

Ladrière, Jean, 132–34
Lamarck, Jean Baptiste, 70, 79, 82; and *Philosophie Zoologique*, 70
Laplace, Pierre, 58
Lateran Council, Fourth, 54
Lavoisier, Antoine, 55, 178
Leclerc, Georges-Louis. *See* Buffon
LeConte, Joseph, 84
Lemaître, Georges, 93, 104–6; and Pope Pius XII's endorsement of Big Bang Cosmology, 106
Letter on Sunspots, 43
Letter to the Grand Duchess Christina, 43, 47, 49
Leucippus, 53
Linnaeus, Carl, 66–67; and his *Systema Naturae*, 69
Lonergan, Bernard, 139–41
Louvain, University of, 55
Love, 163–64
Lucretius, 53–54, 67
Lumen Gentium, 149
Luther, Martin, 25–27
Lyell, Charles, and his *Principles of Geology*, 62, 69

Magic, 55
Magisterium, 29, 85
Manicheism, 8–9
Marburg Colloquy, 28
Marriage, 202–3
Martyr, Justin, 7
Materialist reductionism, 58, 77
Mathematics, 16, 35, 43, 46, 50; Laplace and, 58; and the mechanical philosophy, 52
Maximillian, Emperor, 27
McMullin, Ernan, 134–37

Index

Mécanique Céleste, 58
Mechanical philosophy, 50, 52, 54; extended into biology, 67
Mersenne, Marin, 49, 54
Mendel, Gregor, 86–87
Miller, Kenneth, 126
Mirandola, Pico della, 60
Mivart, St. George Jackson, 73
Modernism, 80, 88
Molloy, Gerald, 69
Monasticism, 12–14; positive valuation of physical labor, 13; and transmission of science, 13
Monogenism, 102
Mosaic account, 68, 84, 88

Narratio prima, 35
Natural history, 2–3, 62–65; historicization of, 68–71, 80; professionalization of, 66–67
Natural law, 151, 155, 161
Natural philosophy, 2–3, 52
Natural theology, 50, 53, 56
Nebular hypothesis, of Laplace, 58
Neo-Darwinian synthesis, 82, 88–89
Neoplatonism, 15–16, 19, 22, 39
Neo-Scholasticism, 76, 80
Newman, John Henry Cardinal, 71–73, 177–81; his *The Idea of a University*, 71; text, 177–81
Newton, Isaac, 51, 57
Nicea, Council of, 7
Nicene Creed, 7
Nicholas of Cusa, 22
Ninety-Five Theses, 26
Noah's Flood, 64, 68–69, 71; rejection of its universality by Brennan, 84
Noonan, John, 151
Notre Dame University, 80
Nova ("new star"), 38

Oakes, Edward, 144–45
Oresme, Nicole, 21–22, 38
Original sin, 168–69
Orthodox Church, 1–2
Osiander, Andreas, 35

Oxford Movement, 71
Oxford University, 15–16, 71

Pachomius, Abbot, 13
Padua, University of, 43
Paley, William, 62, 65; *Natural Theology*, 65, 70, 78
Paracelsus, 55
Paris, University of, 15, 30
Pascal, Blaise, 49, 60
Pascendi Dominici Gregis, 88
Patristic era, 6–7; and science, 8; transmission of Greek science, 8
Phenomenology, 111, 133
Physico-theology movement, 65–67
Physics, 178–79
Physiologus, 14
Physique de Croyant, 58
Pisa, University of, 41, 43
Plato, 4, 22, 165
Platonism, 19, 39, 46
Pliny the Elder, 5, 64
Polanyi, Michael, 137
Polygenism, 190
Pontifical Academy of Life, The, 158, 162
Pontifical Academy of Sciences, The, 39, 55–57, 103
Pontifical Biblical Commission, The, 99
Pope Benedict XIV, 58
Pope Benedict XVI, 147
Pope Clement IV, 16
Pope Clement VII, 28
Pope Gregory XIII, 40, 56
Pope John XXIII, 92
Pope John Paul II, 114, 191–99; and *Centesimus Annus*, 156; and the 1996 statement to the Pontifical Academy of Sciences dealing with evolution, 121, 165, 196–99
Pope Leo I, 7
Pope Leo X, 27
Pope Leo XIII, 56, 75, 96
Pope Paul III, 25, 28–29, 39
Pope Paul IV, 30
Pope Paul VI, 113–14, 150

Pope Pius VII, 55
Pope Pius IX, 55, 74, 78
Pope Pius XI, 57, 103
Pope Pius XII, 76, 93, 97–98, 100–106, 187–90; and Big Bang cosmology, 105–6
Pope Urban VIII, 45, 48
Pope, Stephen, 164
Potentia Dei, absoluta and *ordinata*, 21
Printing, 64
Probabilism, Jesuit theory of, 50
Process thought, 137
Protestantism, 47, 50
Protestant Reformation, 26–27
Providentissimus Deus, 75, 88
Psalms, 4
Ptolemaic cosmology, 22, 30, 33, 35–36, 38, 45, 59
Ptolemy, 5
Pusey, E. B., 73
Pythagoras, 4

Quadrivium, 15, 17, 47
Quantum theory, 96

Rationes seminales, Aristotelian and Stoic idea, 63; Mivart's *logikoi spermatikoi*, 77
Ray, John, 65, 69; *The Wisdom of God Manifested in the Works of Creation*, 65
Realism, 121
Reason, 57
Reformation, 2, 25–30
Relativity theory, 96, 104
Renaissance, 39–40, 46; "naturalism" in, 52
Retroduction, 134, 136–37
Retrograde motion, 36
Rheticus, Georg Joachim, 35, 60
Riccioli, Giovanni Battista, 51–52, 59
Roman thought, 5
Rome: collapse of empire, 12; Catholic Church in, 47–48
Royal Society of London, 64, 76
Rule of Saint Benedict, 13

Saint-Hilaire, Étienne Geoffroy, 70
Sapientia, 2
Scarpellini, Feliciano, 55
Schmalkaldic League, 27
Scholasticism, 3, 15–23, 39–40, 46, 52, 80
Schönborn, Christophe Cardinal, 147
Scientia, 2
Scientific Revolution, 25–26, 35–38, 50, 52–53
Scientist, first used, 3
Scriptural geology, 71
Secchi, Angelo, 56
Second Vatican Council. *See* Vatican II
Secundum imaginationem, thought experiments, 21
Septuagint (Greek New Testament), 68
Sergeant, John, 53–54
Sex, 151; and Pope John Paul II's theology of the body, 152–53
Sidereus nuncius, 43
Siger of Brabant, 19
Skepticism, 50
Social teaching (Catholic), 156
Sociobiology and human behavior, 163–65
Sola scriptura, principle of, 28–29
Sophia, 2
Sorignet, Abbé, 69
Soul, 5; Mivart's "protective belt" around, 76, 165
Spalding, Bishop John L., 85–86
Spencer, Herbert, 83, 163–64
Stellar parallax, 43, 46
Stem cell research, 159
Steno, Nicholas, 49, 64
Stoeger, William, 124–26
Summa Theologiae, 15, 17
Syllabus of Errors, 74

Tartaglia, Niccoló, 52
Taxonomic nomenclature, 66
Teilhard de Chardin, Pierre, 89–90, 93, 100, 106–12, 137; and Christ, 110, 112; de Lubac's study of, 109; and the *monitum*, 109; and 'Piltdown Man,' 109–10

Teleology: Aristotle and, 5; Duhem and, 59; in physico-theology, 65, 67
Telescope, first used astronomically, 43
Tempier, Stephen, 19
Tertullian, 7–8
Tetzel, Johann, 26
Theism, 65, 76–77, 81
Theodoric of Freiburg, 22
Theodosius, Emperor, 6
Theology, as "Queen of the Sciences", 47
Topsell, Edward, and *History of Foure-Footed Beasts*, 64
Torricelli, Evangelista, 53
Transubstantiation: Eucharistic theology of, 46, 54; Laplace's contempt for, 58
Trent, Council of, 25, 27–28, 30, 40, 47, 50
Tridentine Catholicism, 50
Trinity, 7
Trivium, 15, 17
"Two Books" metaphor, 70
"Two languages" model, 58, 71
Tychonic geo-heliocentric model, 36

Ultramontanism, 74
Uniformitarianism, in geology, 62, 69
Unitas scientiae, 23, 41, 49, 51, 72

Universities, foundation of, 15, 23
Ussher, Archbishop James, 68

Vacuum, 52, 53
Valla, Lorenzo, 39
Van Helmont, Jan, 55
Van Leeuwenhoek, Anton, 64
Vatican I, Council, 74
Vatican II, Council, 28, 91–93, 97, 112–14, 117–20, 149
Vatican Observatory, 40, 56, 124
Venus, phases of, 46
Von Haller, Albrecht, 67

Wallace, William, 95
Walworth, Clarence, 69
Werner, Abraham, 70
Whiston, William, 65
White, Gilbert, and *The Natural History and Antiquities of Selborne*, 65
White, Lynn, 156
Whitehead, A. N., 137–38, 146
Windle, Bertram, 98
Wiseman, Nicholas Patrick, Cardinal, 69
Wittgenstein, Ludwig, 132–34
Woods, Henry, 89

Zahm, John Augustine, 80–85; *Evolution and Dogma*, 80; text, 182–87

About the Authors

PETER M.J. HESS is an adjunct professor at Saint Mary's College. Until recently he was the Associate Program Director of the Science and Religion Course Program, Center for Theology and the Natural Sciences at the University of California, Berkeley. He is the author of numerous book reviews and articles on science and Catholicism. He is on the editorial board of the series.

PAUL L. ALLEN is Assistant Professor in Theological Studies at Concordia University in Montreal, Quebec.